FLEXIBLE URE

Related books

CULLINANE & STOKES	Rural Transport Policy
HENSHER (ed.)	Travel Behaviour Research: The Leading Edge
HENSHER & BUTTON (eds.)	Handbooks in Transport Series
MAHMASSANI	In Perpetual Motion: Travel Behaviour Research Opportunities and Application Challenges
ROOT	Delivering Sustainable Transport: A Social Science Perspective
TANIGUCHI	City Logistics
TANIGUCHI & THOMPSON	Logistics Systems for Sustainable Cities
TURRÓ	Going trans-European: Planning and Financing Transport Networks for Europe

Related Journals

Transport Policy;
Editor: M. Ben-Akiva

International Journal of Transport Management:
Editor: M. McDonald

Transportation Research A : Policy and Practice
Editor: F.A. Haight

Transportation Research B: Methodological
Editor: F.A. Haight

Transportation Research C: Emerging Technologies
Editor: S. Ritchie

Transportation Research D; Transport and Environment
Editor: K.Button

Transportation Research E: Logistics and Transportation Review
Editor: W. Talley

Transportation Research F: Traffic Psychology and Behaviour
Editors: J. Rothengatter and J. Groeger

Cities
Editor: A. Kirby

Land Use Policy
Editor: A. Mather

FLEXIBLE URBAN TRANSPORTATION

By

Jonathan L Gifford
George Mason University, Arlington, USA

2003

An Imprint of Elsevier Science

Amsterdam - Boston - London - New York   Oxford - Paris - San Diego
San Francisco - Singapore - Sydney - Tokyo

ELSEVIER SCIENCE Ltd
The Boulevard, Langford Lane
Kidlington, Oxford OX5 1GB, UK

© 2003 Elsevier Science Ltd. All rights reserved.

This work is protected under copyright by Elsevier Science, and the following terms and conditions apply to its use:

Photocopying
Single photocopies of single chapters may be made for personal use as allowed by national copyright laws. Permission of the Publisher and payment of a fee is required for all other photocopying, including multiple or systematic copying, copying for advertising or promotional purposes, resale, and all forms of document delivery. Special rates are available for educational institutions that wish to make photocopies for non-profit educational classroom use.

Permissions may be sought directly from Elsevier's Science & Technology Rights Department in Oxford, UK: phone: (+44) 1865 843830, fax: (+44) 1865 853333, e-mail: permissions@elsevier.com. You may also complete your request on-line via the Elsevier Science homepage (http://www.elsevier.com), by selecting 'Customer Support' and then 'Obtaining Permissions'.

In the USA, users may clear permissions and make payments through the Copyright Clearance Center, Inc., 222 Rosewood Drive, Danvers, MA 01923, USA; phone: (+1) (978) 7508400, fax: (+1) (978) 7504744, and in the UK through the Copyright Licensing Agency Rapid Clearance Service (CLARCS), 90 Tottenham Court Road, London W1P 0LP, UK; phone: (+44) 207 631 5555; fax: (+44) 207 631 5500. Other countries may have a local reprographic rights agency for payments.

Derivative Works
Tables of contents may be reproduced for internal circulation, but permission of Elsevier Science is required for external resale or distribution of such material.
Permission of the Publisher is required for all other derivative works, including compilations and translations.

Electronic Storage or Usage
Permission of the Publisher is required to store or use electronically any material contained in this work, including any chapter or part of a chapter.

Except as outlined above, no part of this work may be reproduced, stored in a retrieval system or transmitted in any form or by any means, electronic, mechanical, photocopying, recording or otherwise, without prior written permission of the Publisher.
Address permissions requests to: Elsevier's Science & Technology Rights Department, at the phone, fax and e-mail addresses noted above.

Notice
No responsibility is assumed by the Publisher for any injury and/or damage to persons or property as a matter of products liability, negligence or otherwise, or from any use or operation of any methods, products, instructions or ideas contained in the material herein. Because of rapid advances in the medical sciences, in particular, independent verification of diagnoses and drug dosages should be made.

First edition 2003

Library of Congress Cataloging in Publication Data. A catalog record from the Library of Congress has been applied for.

British Library Cataloguing in Publication Data. A catalogue record from the British Library has been applied for.

ISBN: 0-08-044053-3

⊚ The paper used in this publication meets the requirements of ANSI/NISO Z39.48-1992 (Permanence of Paper). Printed in The Netherlands.

The illustration on page 19 has been reproduced with permission from Walters Art Gallery, Walters Art Museum, Baltimore.

Contents

	Preface	ix
1	**Transportation and the Economic Vitality of Communities**	1
	The Urban Transportation Dilemma	2
	Transportation Planning in Context	6
	Infrastructure and economic growth	7
	Structural changes	9
	Changing government roles and responsibilities	11
	The Need for a New Approach	13
2	**Order, Efficiency, and the Struggle against Chaos**	17
	Order and Efficiency in the Italian Renaissance	17
	Chaos and the Search for Order	19
	The American Renaissance	21
	The Progressive Era	22
	Scientific management	23
	Standardization	23
	Public administration	25
	City Planning and the City Beautiful	26
3	**The American Highway Program to 1956**	29
	The First-Generation Highway Program	30
	The 1920s	31
	The early Depression: 1930-1936	33
	The late Depression: 1937-1941	34
	Planning the Interstate System	35
	Toll roads versus free roads	37
	Urban versus interregional highways	39
	Design standards	40
	Limited mileage	43
	Cost estimates, financing, and the highway trust fund	44
	Beltways	44
	Institutional arrangements	46
4	**The American Highway Program since 1956**	49
	Building the Interstate: 1956-1991	50
	The 1956 Highway Acts	50
	The "Golden Age"	51

	The Freeway Revolt	53
	Process, design, and values	55
	Interstate 66, Fairfax and Arlington Counties, Virginia	56
	Interstate 70, Glenwood Canyon, Colorado	60
	Franconia Notch, New Hampshire	63
	"Finishing" the Interstate	65
	The Post-Interstate Era	66
	The Intermodal Surface Transportation Efficiency Act of 1991 (ISTEA)	69
	National Highway System Designation Act of 1995	74
	Transportation Equity Act for the 21st Century (TEA-21)	75
5	**Transportation Planning Methods**	**79**
	The Institutional Context of Transportation Planning	80
	The Role of Analysis in Transportation Planning	83
	Decision-Making Concepts	85
	Indicative planning	85
	Margin of safety analysis	87
	Benefit–cost analysis	88
	Distributional equity	89
	Sources of uncertainty	91
	Irreversibility, non-pecuniary impacts, and other limitations	95
	Probabilistic risk assessment	96
	Life-cycle cost analysis (LCCA)	98
	Least cost planning	99
	Investing under uncertainty	100
	Varying construction cost	102
	Varying demand	103
	Increasing uncertainty over demand	104
	"Bad News Principle"	104
	Scale versus flexibility	106
	Environmental applications	107
	Investing under uncertainty: summary	107
	Participatory decision making	108
6	**The Evolution of Transportation Planning**	**111**
	Highway Planning Prior to World War II	111
	Metropolitan Transportation Planning	112
	Urban travel modeling and the Interstate program	118
	National Environmental Policy Act	118
	Alternatives analysis	119
	ISTEA	119

		vii
	Air Quality Planning	122
	Period I: 1970-1990	122
	Period II: the 1990s	124
	Transportation planning and the control of air quality	128
	National-Scale Planning	130
	Highway needs studies	131
	Condition and performance reports	134
	Economic productivity studies	137
7	**Challenges to the Neoclassical Economic Paradigm: Complexity, Adaptation, and Flexibility**	**139**
	Increasing Returns	140
	Sources of increasing returns	142
	Properties of increasing returns systems	147
	The Austrian School	151
	Choice under uncertainty	152
	Competition as learning	153
	Capital structure	154
	Implications for Infrastructure Planning	156
8	**The Need for a New Approach**	**159**
	Four Fallacies of the Current Approach	160
	The exogenous goal fallacy	160
	The predictive modeling fallacy	161
	The efficiency fallacy	163
	The public involvement fallacy	163
	A Reinvented Transportation Planning Process	164
	Control	164
	Flexibility	165
	Control and flexibility	167
	Adaptive discovery	169
	Chapter Appendix: Bibliography on Concept of Flexibility	173
9	**Transportation Planning: A Flexible Approach for the Twenty-First Century**	**175**
	Stability and Agility	176
	Reinventing Transportation Planning	177
	A commitment to honesty	177
	Intelligence: a source of factual information	180
	Decision support	182
	Design and implementation	184
	Design and procurement	185

	Outreach and community involvement	186
	Monitoring	188
	Performance indicators	189
	Monitoring flexibility	190
	Assessing financial and economic viability	194

10 Reality Check: Institutionalizing Flexible Transportation Planning — 197

Going Cold Turkey	198
The National Campaign	198
Consortia and Informal Voluntary Organizations	199
TRANSCOM	201
E-ZPass	201
The Data Sharing Model	203
A Local, State, and Regional Implementation Strategy	205
Metropolitan planning organizations	205
Informal voluntary organizations	206
The role of the states	207
Flexible Planning at the Federal Level	208
Identify and address uncertainty in program plans and analyses	208
Shorten the time horizon for project analysis	211
Adopt incentives based on measurable outcomes	212
Facilitate sharing of information	214

11 An Agenda for Action — 217

Local, State, and Regional Actions	217
Establishment of an intelligence function	218
Develop a decision support function	219
Design and implementation	220
Conduct Monitoring	221
National Actions	221
Deregulate the metropolitan transportation planning process	222
Revise federal policies to focus on measurable outcomes	222
Streamline the environmental approval process	222
Shorten the time for project cost–benefit analysis	223
Support a national or regional data interchange standards process	224
Reorient national planning around macroeconomic analysis	225

12 Conclusion — 227

Index — 233

Preface

This book is a critique of transportation planning as it is practiced in the United States today and a proposal for a new, more flexible approach. The U.S. is now facing profound challenges to its economic competitiveness and social equity, to public safety and security, and to the integrity of its environment. The ability to create transportation systems that contribute to addressing those challenges effectively requires a planning process radically different from the process in place today. Meeting the nation's challenges effectively requires flexibility, honesty about what does and does not work, transparency, and inclusion of a broad range of stakeholders. The current process is rigid, dishonest—the process, that is, not the professionals who work in it—opaque, and exclusive.

This call for reform is in some ways both naïve and imperfect. The current transportation planning process is deeply ingrained in institutions and procedures that direct substantial funds to well-entrenched interests. It is unlikely that a new approach can displace the status quo any time soon, and any real change in practice that the proposed reform might engender will invariably raise questions that the book fails to address. Nonetheless, I offer it in the spirit of constructive criticism on a matter of great societal urgency.

The motivation for writing this book goes back two decades to my doctoral dissertation on the planning and design of the interstate highway system and its impacts on American cities. In that work I asked, how could a program as widely welcomed and well-intentioned as the Interstate program in 1956 have unleashed such a furious rejection in so many cities only a decade later? The answer lay in the nature of bureaucratic politics and the perils of implementation and unintended consequences.

This book takes the logical next step and examines the difficult and humbling question of what can and should be done to remedy the transportation planning crisis. The book describes how transportation planning has reached its troubled present state, and prescribes a way forward. Many of the ideas and proposals presented here are not wholly new. Indeed, the proposed approach builds on what is best about transportation planning today. It seeks to relax some of the procedural and societal constraints on discovering the proper balance between transportation improvements and other objectives of the society those improvements are intended to serve. Yet while promising signs of improvement are apparent here and there, much about the current practice of transportation planning reflects the best thinking of the 1950s, frozen in the amber of regulations, consent decrees, and procedural checkpoints. These frustrate attempts for reform, with the consequence that the transportation system fails to serve society as well as it could.

The book has been a labor of many years, and I owe a debt of gratitude to many. At George Mason University, Louise White helped inspire the writing. Jim Pfiffner offered valuable advice and counsel. Roger Stough and the School of Public Policy provided extremely generous encourage ment and financial support. Graduate students Sanjay Marwah, Danilo Pelletiere, and Odd Stalebrink have provided indispensable research assistance. For several years, Mary Clark provided essential administrative support, as well as great working companionship. Many colleagues debated and discussed ideas presented in the book. And the university itself provided an intellectual setting in which I could develop and complete the manuscript.

Outside the university, I am indebted to my editor, Chris Pringle and his able and patient staff at Elsevier Science, to Richard Rowson, who provided invaluable editorial advice, to Catherine Kreyche, who copyedited the manuscript and supervised its preparation, and to Thanigai Tiruchengodu for his assistance with computer graphics. Last but not least, my good friend Bob Vastine provided warmly appreciated support, prodding, and encouragement over the long course of its development. To all, a sincere and heartfelt thank you.

Of course, the normal disclaimers apply.

1

TRANSPORTATION AND THE ECONOMIC VITALITY OF COMMUNITIES

A century and a half ago, in 1847, the author's ancestors fled the poverty, harsh climate, and rocky terrain of their native Norway and settled in central Iowa, twenty miles north of Des Moines—a land, they found, "flowing with milk and honey." A few decades later, in 1874, came a narrow-gauge railroad, which began service between Des Moines and the town of Ames, fifteen miles north. The settlers established a town next to the rail line and called it Sheldahl, after Osmund Sheldahl, the author's great grandfather, who had donated the land. Five years later the Northwestern Company purchased the line and upgraded it to standard gauge. Sheldahl prospered, so much so that at one point it even supported eleven saloons.

Another five years later the Chicago, Milwaukee, and St. Paul Company announced plans to build an east–west line that would pass only a quarter mile north of the center of Sheldahl. Competition with the Northwestern, the town hoped, would bring more favorable rates. But the planned route required two river crossings and traversed some difficult terrain. New surveys identified a more favorable route, but it passed one and one-half miles north of Sheldahl. The town sent an emissary to the railroad to advocate the original route, but all he got for his efforts was a free ticket home.

A remarkable thing happened after service began on the new line. The residents and merchants of Sheldahl literally picked up their town and moved it. They laid a trail across the prairie and moved more than fifty buildings to the new crossing. Each building was jacked onto wheeled "trucks" and pulled with the aid of circular horse-power. Immediately ahead of the building itself the trail was planked with heavy boards, which were continuously resupplied from the rear. It was slow work, often requiring a week or even ten days for one building. The horse-powered apparatus, while it required frequent stops for resetting, provided tremendous mechanical advantage, allowing many of the larger buildings, including the grain elevator, to be moved the full two miles with a single horse.

2 FLEXIBLE URBAN TRANSPORTATION

In the end, Sheldahl lost more than half of its 347 residents. Many of those who stayed behind harbored hard feelings against those who left. But Slater, as the new town was eventually called, prospered and today has one of the largest grain elevators in the state.[1]

The ten years from boom to bust in Sheldahl are a poignant example of how transportation infrastructure affects the economic vitality of communities. As in Sheldahl, transportation infrastructure is a powerful determinant of the economic and social well being of all cities, towns, and communities. Transportation infrastructure provides access for companies, factories, and farms to the work force and supplies they need to produce their products and services and distribute them to their customers. And it provides access to work, church, shopping, and recreation for families.

These impacts add up from city to city and town to town such that transportation infrastructure decisions are important determinants of social well being at a national level as well. Seldom are the economic implications of a particular decision as stark as they were for Sheldahl in 1884. But when transportation infrastructure is at cross-purposes with the needs of a community, economic and social well being can suffer.

The nature of the relationship between infrastructure and quality of life is complex, however. The demand for transportation infrastructure arises out of private and collective decisions by households, firms, and units of government whose motivations are not always well or easily understood by those who plan transportation infrastructure. Moreover, transportation infrastructure arises out of decision processes, largely in the public sector, that are influenced not only by technical and engineering considerations but also by the harsh tug of partisan, parochial politics.

It is not surprising, then, that "disconnects" occasionally arise between transportation infrastructure (the supply side) and community needs (the demand side). Indeed, what may be surprising in an era of skepticism about politics and political institutions is that the system works as well as it does.

This book is about the growing disparity between the supply and demand for transportation infrastructure, its consequences for social well being, and proposals for change that would bring facilities and demand into closer accord.

THE URBAN TRANSPORTATION DILEMMA

Transportation in most American cities reflects a dilemma. On the one hand, traffic demand has increased dramatically over the last quarter century due to prosperity and population growth. The supply of highways, on the other hand, has in most places grown only minimally.

---

[1] Don Fatka, "From Norway to Story County" (January 1970); and James A. Storing, "The Town That Moved," *The Palimpsest* (State Historical Society of Iowa, February, 1939), cited in "A Town Is Formed." Both items in "History Book: Sheldahl, Iowa," ms.

The detailed reasons for the stagnation in new supply vary from place to place. But at a general level, it is fair to say that proposals to build new or expand existing highways fail the test of implementation; they somehow fail to muster sufficient support to overcome the costs and barriers to implementation.

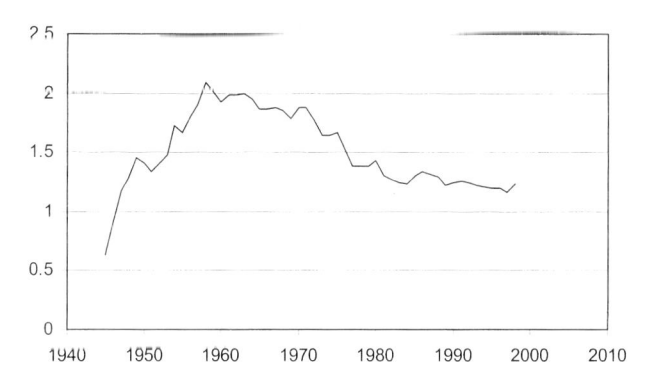

Figure 1-1
U.S. Disbursements for Highways as a Percentage of GDP: 1945-1998

Source: GDP: U.S. Bureau of Economic Analysis <http://www.bea.doc.gov>;
highway disbursements: U.S. Federal Highway Administration, *Highway Statistics* (various years).

The public consensus for expanding and improving urban transportation infrastructure has eroded significantly in the last several decades. Its erosion is evident in the opposition to new and expanded urban highway projects, despite worsening congestion and seemingly inexorable increases in public demand for automobiles and road space. Its erosion is also evident in flat or declining budgets for highway construction and maintenance and in the deferred maintenance of existing facilities (Figure 1-1).

Why is this a dilemma? Increasing traffic without commensurate increases in supply almost necessarily increases the congestion and delays that travelers face. "Excess capacity" existed in some places twenty five years ago, even at peak hours. And most highway facilities and transit services have excess capacity between 10 p.m. and 6 a.m. But transportation demand exhibits strong variation by time of day because travel is a means to participate in other social and economic activities that also ebb and flow by time of day. Most areas have long since used up any excess capacity that existed at peak hours so that increased demand is accommodated by spreading the peak to longer and longer periods.

If travel confers net benefits on the traveler (and who but the traveler is in a position to dispute that assessment?), then society is better off unless the costs not borne by travelers, such as noise and air quality, are so great that they outweigh the benefit of the trip to those traveling. (Congestion costs are borne by other travelers and hence are internal to the travelers' decisions.) But the assertion that it is impossible to build one's way out of congestion misses the point. The presence of congestion indicates that the value of travel afforded by a facility or service exceeds the costs to the traveler of making the trip, including dealing with congestion.

A larger question is whether a mobile society is better off than an immobile society. Clearly it is, to a significant degree. Mobility confers choice from a range of spatially dispersed activities. The cheaper, faster, and easier it is to travel, the greater the field of activities available.

Public opposition to large infrastructure projects is nothing new. The construction of the Erie Canal in the early nineteenth century precipitated thousands of complaints and lawsuits claiming damages suffered as a result of its construction and operation.[2] What is new is that public opposition today is more often successful in stopping, delaying, or modifying the design of many major projects.

Building urban transport infrastructure facilities requires power—power to condemn and take private property, power to award and enforce franchises and contracts, and power to tax or levy tolls. The construction of the grand boulevards of Paris between 1852 and 1870 required widespread demolition to open large corridors and achieve a harmonious and technically efficient configuration of streets and vistas. Baron Haussmann could only execute such a plan with the power and authority of Napoleon III behind him.[3] In the U.S. between the 1920s and the 1960s, the man responsible for the construction of the lion's share of greater New York City's public facilities was "power broker" Robert Moses.[4]

In a free society, the exercise of such power requires public support. There have been times when the public supported major infrastructure improvements: times of crisis, times of overwhelming public demand, times of widespread belief in the solutions being offered. But that support is less often present today. The question is why, and what to do about it?

If travel confers such great benefits, why has expanding the capacity to travel through the provision of highways and other facilities and services so often failed to pass the test of implementation? One explanation is that there is no point in building new supply because traffic simply expands to fill it up. It has become almost axiomatic that, "You can't build your way out of congestion." This is in many respects a wrong-headed view. People travel because they find it productive to do so. To be sure, there is some recreational travel, that is, situations where the travel is an end in itself. And recreational travel in some locales causes congestion. (Consider the congestion of cars cruising Main Street in small towns on a Saturday night.) But generally speaking, people travel because they find that the cost and inconvenience imposed by travel-

---

[2] Carol Sheriff, *The Artificial River: The Erie Canal and the Paradox of Progress, 1817-1862* (New York: Hill & Wang, 1996).
[3] Howard Saalman, *Haussmann: Paris Transformed*, Planning and Cities (New York: Braziller, 1971).
[4] Robert A. Caro, *The Power Broker* (New York: Knopf, 1974).

ing are more than offset by the benefits conferred by participating in the activities at the end of their trips.

Another explanation for the decline in broad public support for urban transport infrastructure is increasing public concern about social and environmental impacts. The catalog of social and environmental costs is extensive, ranging from poor air quality (local, regional, and interregional), to land use problems, to community separation, to (for highways) erosion of viable markets for public transit, to military interventions in the Persian Gulf.[5] Growing public concern about these issues could derive from a number of factors. One factor could be that appreciation for the environment and habitat may be "income elastic," that is, as incomes rise, the value individuals place on non-subsistence amenities grows. Another factor could be that public awareness of these costs has grown over the last several decades, and public willingness to accept them has declined. Yet another factor could be that environmental elitists have "hijacked" decision making and seek to use congestion to force Americans to give up their cherished lifestyle, which is so energy and land intensive.

A third and related explanation for the erosion of public support is that the demand for highways reflects large direct and indirect subsidies that have biased transport users in their favor. Such subsidies include dedicated gas taxes, general fund support, as well as a number of services such as policing and street cleaning that are not typically charged to the highway account. These subsidies have caused highways to be priced below cost, according to this line of reasoning, and consumers have rationally acted to exploit the subsidies, driving the demand for highways above economically efficient levels. A related point is that the congestion costs of driving during peak periods are external to the traveler, leading to over-consumption of peak period travel.

A fourth common explanation is a decline in public trust in the institutions charged with building urban transport infrastructure, and a commensurate refusal to accede to their decisions. The deference to technical expertise that once shielded decisions from public challenge and scrutiny has eroded. Instead, society defers to sentiments of "not in my backyard" (NIMBY), or more emphatic sentiments of "build absolutely nothing anywhere near anyone" (BANANA). Procedural requirements for environmental assessments and public participation have led to what one author has termed "demosclerosis," the inability of a democratic society to act decisively on important but divisive issues.[6]

Some allege that the real culprit is the perversity of the American people. They want all the benefits of mobility without the cost and inconvenience of building new facilities, and they refuse to confront the patterns of land development, vehicle ownership, and usage that give rise to traffic growth.

---

[5] Mark A. Delucchi, *The Annualized Social Cost of Motor-Vehicle Use in the United States, 1990-1991: Summary of Theory, Methods, Data and Results*, report UCD-ITS-RR-96-3 (1) (Davis: University of California, Davis, Institute of Transportation Studies, June 1997).
[6] Jonathan Rauch, *Demosclerosis: The Silent Killer of American Government* (New York: Times Books, 1994).

While there is some truth in each of these explanations, they all tend to place the responsibility for eroding public support outside the transportation supply system. This book argues that a large part of the blame lies not with shortsighted politicians and ignorant, irrational voters but with the infrastructure community itself. Infrastructure suppliers have simply failed to deliver facilities that are publicly acceptable.

The erosion of public consensus for urban transportation infrastructure reflects public dissatisfaction with the facilities being tendered for consideration. What urban highway suppliers have been providing has not offered the combination of operational features and environmental and community impacts that the public is willing to accept. Suppliers have failed to refine and adapt their designs to provide features and services that users are willing to support. Absent such support, political decision makers have withdrawn financial and political resources from the institutions that supply infrastructure.

Out-of-scale, poorly conceived, and insensitively implemented projects, especially under the auspices of the Interstate program, caused a backlash against freeways that has hampered progress on many highway projects for the last three decades. In short, the infrastructure community ignored its customers, and the consequences have been enormous. The corollary of this observation is that the infrastructure community can again bring forth projects and ideas that capture the public's imagination.

It would be a mistake to conclude that the public rejects urban highways and favors more public transit. Public behavior clearly refutes that conclusion. With few exceptions, market share is down on all modes of transport other than driving alone. The public is more than willing to use what suppliers have provided. What alternative do they have? Staying at home? But as urban transport projects—especially urban freeways—have been deployed in urban areas, the public reaction in many instances has been firm and decisive opposition.

The problem, then, is a failure of infrastructure suppliers to conceptualize and design facilities that command widespread public support. The remedy must be to discover what kinds of facilities the public will support.

The time may well be right for a new approach. Public support may have reached its nadir in the 1990s and begun to recover. Congress has recently increased transportation funding significantly, and will soon be considering reauthorization of the federal surface transportation program. It is critical not to squander the opportunity by offering up the same old ideas. It is essential to embrace a customer orientation, to discover what customers value and will support.

TRANSPORTATION PLANNING IN CONTEXT

The current state of urban transportation planning is the product of a century-long tradition. It reflects the evolution of thinking about the sources of economic growth, the structure of the economy, and the role of government.

Infrastructure and Economic Growth

"Getting the transportation system right" to support growth of the economy is an appealing objective. Poor infrastructure can clearly hurt the economy, even though many societies have maintained high levels of economic output over the short term when their infrastructures have been destroyed by natural disaster or war.[7] But the relationship between infrastructure investment and economic output and productivity has only recently become the matter of formal econometric study.

Society's views about the linkage between infrastructure investment and economic growth and the appropriate role for government in the promotion of infrastructure development have changed considerably in the last two centuries. Soon after the creation of the republic, proposals emerged for federal sponsorship of road and canal improvements intended to improve access across the Appalachian Mountains, where population was growing rapidly. One of the earliest projects was the National Road from Maryland to Ohio, constructed between 1822 and 1838. The National Road had been built with the support of Senators Henry Clay (Kentucky) and John C. Calhoun (South Carolina), who were concerned that introduction of steamboat service on the Mississippi and Ohio rivers would imperil access from east coast cities to (then) growing western markets. But federal support for the National Road and other similar "internal improvements" soon ran afoul of greater concerns about states' rights and limiting the authority of the federal government, and funding ceased in 1838. Road and canal access across the Appalachian Mountains would thereafter proceed as a state-sponsored or state-chartered enterprise.

The state-sponsored Erie Canal connected New York City to Lake Erie and had profound impacts on the spatial distribution of products. When it opened, the cheapest route for bulk commodities from Philadelphia to Pittsburgh was via New York. The canal's stunning success prompted many states to launch their own canal initiatives. But the subsequent failure of other projects like the Chesapeake and Ohio Canal caused a backlash against further state support.

Federal support soon returned with a vengeance, however, to sponsor railroad expansion with land grants, loans, subsidies, and tariff remission on rails. The land grants were by far the most valuable. By one account, the federal government gave 158 million acres of land to the railroads, an area almost the size of Texas. Even after the forfeiture of almost 42 million acres due to failure to fulfill the grant conditions, the total area was still 15 percent larger than the state of California. States, counties, and municipalities also aggressively sponsored railroad projects, including state land grants to railroads of more than 50 million acres, and hundreds of millions of dollars in direct subsidies and subscriptions of railroad stock.[8]

---

[7] Cf. Konvitz's work on war economies and the strategic bombing survey, which concluded that during World War II, the ability of communities to survive serious disruptions to their basic infrastructure services often far exceeded military planners' expectations. Josef W. Konvitz, *The Urban Millennium: The City-Building Process from the Early Middle Ages to the Present* (Carbondale: Southern Illinois University Press, 1985).

[8] See David Haward Bain, *Empire Express: Building the First Transcontinental Railroad* (New York: Viking, 1999). for a discussion of the race for the transcontinental railroad and the importance of federal subsidies. For an

Explicit calculations of the economic returns from such investments played only a minor role in decisions about supporting infrastructure. Rather, the benefits of infrastructure development were taken for granted, and the policy debate revolved around how such expansion should be capitalized and what roles government and the private sector should play. The Interstate highway system, planned and constructed from the late 1930s until the 1980s, may have been the last of these great government public works projects.

Beginning with the Flood Control Act of 1936, decisions about publicly financed projects began to reflect examination of the relative benefits and costs of a project in recognition of competing demands for limited resources. Benefits were enumerated in terms of time savings and cost reductions. With the development of benefit–cost analysis came refinements to assess the environmental impacts of projects and the "external" costs that infrastructure users imposed on society in the form of, for example, noise and tailpipe emissions.[9]

In the 1980s, economists began to examine the linkage between transportation infrastructure investment and the productivity of private investment at the level of the individual firm. Good highway access, for example, can allow a company to use just-in-time inventory management, whereby production inputs arrive at the factory "just in time" to be used on the production line rather than tying up capital while being stored in costly warehouses.[10]

A burst of research into infrastructure investment impacts began in 1989 with a series of papers by Aschauer.[11] This body of research paints a complicated portrait. It provides no clear indication of whether the U.S. has been spending too much or too little on infrastructure in general or highways and transit in particular. Aschauer's 1994 review essay on the topic called not for more investment, but for institutional structures that would permit state and local governments, which own almost all infrastructure capital, to determine their own optimal levels of investment.[12]

More recent work on this subject suggests that the returns from highway capital investment are roughly comparable to the returns from investment in private capital, about 11 percent, although the returns were much higher—on the order of 35 percent—during the 1950s and 1960s

---

older account, and acreage estimates, see S. E. Morison and H. S. Commager, *The Growth of the American Republic*, 1930, 4th ed. (New York: Oxford University Press, 1950), II:112.

[9] For a comprehensive discussion of costs for highway travel, see Delucchi, *The Annualized Social Cost of Motor-Vehicle Use in the United States, 1990-1991: Summary of Theory, Methods, Data and Results*.

[10] Richard R. Mudge, "Assessing Transport's Economic Impact: Approaches and Key Issues," paper presented at the Transportation and Regional Economic Development, November 6-8, 1994, Airlie House, Warrenton, Virginia.

[11] David A. Aschauer, "Is Public Expenditure Productive?" *Journal of Monetary Economics* 23 (1989): 177-200; David A. Aschauer, "Public Investment and Productivity Growth in the Group of Seven," *Economic Perspectives* 13, no. 5 (1989): 17-25; David A. Aschauer, "Does Public Capital Crowd Out Private Capital?" *Journal of Monetary Economics* 24, no. 2 (1989): 171-88.

[12] Edward M. Gramlich, "Infrastructure Investment: A Review Essay," *Journal of Economic Literature* 32 (September 1994): 1176-96.

when the Interstate system was being built. These results suggest that highway investment is about where it should be.[13]

But that conclusion is subject to a number of important caveats. First, these estimates reflect only benefits to producers of goods and services and exclude all benefits to consumers, including timesavings and improved accessibility. While such benefits are real, they legitimately fall outside the scope of the productivity measurement system.

Second, the data on highway capital stocks used to estimate the model is seriously deficient. The model uses the book value of highway capital stock, which is its initial cost minus depreciation. The preferred methods for measuring capital stock are either replacement value or market value, the latter of which reflects the net present value of future services provided by the asset in question. Also, the depreciation schedules used to estimate book value have not been updated for decades.

Third, the model assumes that highway capital stocks are a good proxy for the services that flow from highway investments. But, clearly, widening a congested highway section might yield significantly different benefits than opening a new highway section that connects two previously poorly connected points. The apparent drop in returns from highway investment from the 1960s to the 1970s could either reflect a shift in investment from new construction to widening and rehabilitation or a realization of diminishing returns from highway investment—how much of each is an empirical question.[14]

Whether shortages of transportation infrastructure investments are limiting productivity growth thus remains a question that is difficult to answer. But the challenge of transportation infrastructure investment goes much further than simply increasing spending through existing institutions and programs. For that will not improve productivity if it generates projects that are economically sound but impossible to build for political reasons, or are politically attractive but economically unsound.

Structural Changes

Concurrent with the growing understanding of the relationship between infrastructure and economic growth, U.S. society and its economy are experiencing important shifts. The industrial mix has shifted away from heavy manufacturing towards services, which generate very dif-

---

[13] M. Ishaq Nadiri and Theofanis P. Mamuneas, *Contribution of Highway Capital Infrastructure to Industry and National Productivity Growth*, Prepared for the U.S. Federal Highway Administration, Office of Policy Development, Work Order No. BAT-94-008 (September 1996); M. Ishaq Nadiri and Theofanis P. Mamuneas, "Highway Capital and Productivity Growth," in *Economic Returns from Transportation Investment* (Landsdowne, VA: Eno Transportation Foundation, Inc., 1996), Appendix A. See also Marlon G. Boarnet, "Highways and Economic Productivity: Interpreting Recent Evidence," *Journal of Planning Literature* 11, no. 4 (May 1997): 476-86; Marlon G. Boarnet, "Infrastructure Services and the Productivity of Public Capital: The Case of Streets and Highways," *National Tax Journal* 50, no. 1 (1997): 39-57; Marlon G. Boarnet, "Road Infrastructure, Economic Productivity, and the Need for Highway Finance Reform," *Public Works Management and Policy* 3, no. 4 (April 1999): 289-303.
[14] Arthur Jacoby, personal communication with the author.

ferent transport requirements that are still not fully understood. It is still unclear where the shift towards services will reduce transport demand per unit of economic output and where it will increase it. To be sure, the production of services requires less transport of natural resources. But professional services typically require offices, computers, and telecommunications systems, the production and distribution of which have significant transportation requirements. Moreover, one the most important inputs to services is labor in the form of brain power, which may need to move to an office via car or transit, or communicate with a customer or headquarters from mobile or remote offices.

Industrial organization has also been undergoing profound changes. Many organizations are abandoning formal hierarchies in favor of work teams, flat organizations, and a variety of what can be termed "self-organizational" approaches. The hallmarks of self-organization are the absence of formal rules and procedures to govern work and production. In their place, organizations are implementing performance-based evaluation or management by results.

Of course there are limits to the extent one wants to empower employees and abandon formal procedures. One rogue trader should not be able to bring down a two-hundred-year-old institution like Barings Bank. The key issue is which decisions can be safely and effectively delegated—and many can—and which cannot.

From the standpoint of transportation infrastructure, self-organization manifests itself in two important ways. First, it may affect the way organizations that design, build, operate, and maintain infrastructure organize their own activities. To the extent these supply organizations are able to exploit such innovations, so much the better. Far more important from the standpoint of this book are its impacts on the demand side, that is, on the households and firms that are the major sources of demand for transportation services.

A parallel development in industrial production is a reduction in the product cycle through the implementation of flexible production techniques. Manufacturers are increasingly moving away from producing large batches of standardized products toward producing smaller batches of products that are aimed at more specific market niches. In the extreme case, companies produce one-of-a-kind customized products. The key challenge for management is shifting from the exploitation of economies of scale, which reduce unit costs by producing large batches of uniform products, to exploiting economies of scope, where costs are reduced through the ability to spread fixed costs over a broad range of specific product types, and where sales depend on being the first to market innovative products.

The nature of employment is also undergoing significant shifts. Job changes are increasing in frequency, and relatively few workers remain with one employer for an entire career. Temporary workers, even in manufacturing, are becoming increasingly common. Such workers are attractive to employers because they provide flexibility to add and let go production workers to meet short-term fluctuations in demand for products and services. Employees, too—sometimes highly skilled employees—often desire temporary work that allows them to work when they need to and attend to other matters when they deem it desirable.[15]

Self-employment is also increasingly common. In the Washington, D.C., metropolitan area, for example, 23 percent of the jobs are for the self-employed (contractors, proprietorships, etc.), and only 77 percent are traditional wage and salary jobs. Self-employed jobs are growing 3 to 4 percent per year, while wage and salary jobs are growing only 1 percent per year.[16] Work at multiple or mobile sites is also growing, as workers "telecommute" one or several days per week, or work at client sites or from mobile offices.

As a result, it has become increasingly difficult to anticipate what transportation infrastructure the private sector will need. With flexible manufacturing and self-organization, the cycle time for new manufactured products is shrinking towards zero. Service organizations evolve with striking rapidity. And household structure is shifting. The nature of work, shopping, and leisure is changing rapidly. Self-employed and mobile workers have very different transportation needs than those of traditional commuters. Taken together these structural shifts in the economy pose profound challenges to those seeking to anticipate society's transportation requirements.

Changing Government Roles and Responsibilities

It can be easy for a mature bureaucracy to govern by focusing on how efficiently infrastructure is used rather than on how efficiently public infrastructure serves users. Policy turns towards "managing" the private sector with congestion fees and restrictions on users, rather than providing new and improved capacity. There is an excessive focus on making sure that society behaves in a way that makes infrastructure work efficiently, rather than on making sure that the infrastructure supports an efficient society.

As John Friedman observed almost half a century ago:

> During the early stages of a planning agency we usually encounter an air of spirited enthusiasm, of experiment and innovation. Imaginative, creative people are recruited into the ranks of the organization. It appeals to them, for it holds the promise of receptiveness to new ideas. The grooves are not yet cut. Purposes are still fluid, waiting to be shaped.... In time, the agency will become less and less flexible. It will develop an ideology drawn from its own experience and buttressed by the need to defend itself against its enemies.... Self-preservation will become its primary aim, and security will be found in its past record of success.... As the agency develops its own traditions and habits of procedure, a tight network of internal and personal relationships becomes built up. Since this network may be dislocated by innovations, it constitutes an automatic defense against them. Moreover, the agency will start to look to certain "outside" interests

---

[15] Louis Uchitelle, "Temporary Workers Are on the Increase in Nation's Factories," *New York Times*, July 6, 1993, A1, D2.
[16] Stephen Fuller, personal communication with the author.

for political support. To keep this support, it must try not to alienate its friends by pursuing a course that might run counter to their interests. In brief, the agency will become less capable of dreaming big dreams, of exploring new solutions, and of influencing the wants of people in the community where the "poverty of aspirations" limits the horizon of what is thought to be possible.[17]

The federal, state, and local governments, which play an important role in the organization and management of transportation facilities and services, have begun to confront profound changes in their roles and responsibilities. The 1990s unleashed an explosion of interest in and experience with management and administrative reform in government agencies. At the federal level these efforts were evident in the Clinton-era National Performance Review and Reinventing Government initiatives and in the Government Performance and Results Act of 1993.[18] Public sector reforms of a similar character are also appearing in Korea, the United Kingdom, New Zealand, Canada, France, Brazil, Australia, and Sweden. A common theme of these reforms is the use of the market as a model for political and administrative relationships, relying on the theories of public choice, principal-agent, and transaction cost economics.[19]

One of the central themes of the administrative reform movement is attention to results and performance. For many cultural and historic reasons, government agencies have traditionally given great weight to procedural integrity, often at the expense of results and performance. Procurement procedures emphasize fairness and objectivity in making awards, often frustrating the desire of those charged with delivering services for speedy, technically sound implementation.[20] Budget and accounting procedures use "fund accounting," which, unlike the cash and accrual accounting used in the private sector, emphasize accountability for ensuring that funds are spent at the time and for the purposes for which they were budgeted, but give little attention to results or performance.[21] Such procedural restrictions can severely limit a worker's ability to recognize and exploit opportunities and implement necessary changes.[22]

The move towards self-organization also poses a dilemma for governments. Much of the management literature recommends decentralized, team-based organization and management that allows managers to exercise judgment and discretion. But governments are quintessentially

---

[17] John Friedman, "Planning, Progress and Social Values," *Diogenes* 17 (Spring 1957): 98-111.
[18] "Government Performance and Results Act of 1993," P.L. 103-62, *Stat.* 107 (1993): 285.
[19] Linda Kaboolian, "The New Public Management: Challenging the Boundaries of the Management vs. Administration Debate," *Public Administration Review* 58, no. 3 (May/June 1998): 189-90; Jack H. Nagel, guest editor, "The New Public Management in New Zealand and Beyond," special issue, *Journal of Policy Analysis and Management* 16, no. 3 (Summer 1997); Organization for Economic Cooperation and Development, *Governance in Transition: Public Management Reforms in OECD Countries* (Paris, 1995); Organization for Economic Cooperation and Development, *Public Management Developments: Update 1994* (Paris, 1995); C. M. Walton, *Emerging Models for Delivering Transportation Programs and Services: A Report of the Transportation Agency Organization and Management Scan Tour*, NCHRP Research Results Digest, no. 236 (Washington, D.C.: National Research Council, Transportation Research Board, 1999).
[20] Steven Kelman, *Procurement and Public Management* (Washington, D.C.: AEI Press, 1990); Steven Kelman, "White House-Initiated Management Reform: Implementing Federal Procurement Reform," in *The Managerial Presidency*, ed. James P. Pfiffner, 2nd ed. (College Station: Texas A&M University Press, 1999), 239-64.
[21] Andrew C. Lemer, "Building Public Works Infrastructure Management Systems for Achieving High Return on Public Assets," *Public Works Management and Policy* 3, no. 3 (January 1999): 255-72.
[22] Israel M. Kirzner, *Competition and Entrepreneurship* (Chicago: University of Chicago Press, 1973).

formal organizations, subject to strict rules and procedures. It is difficult for such institutions to adapt to highly dynamic environments that are not subject to predictability and control.[23]

How and whether American government can adopt self-organization approaches will continue to be a question of profound importance. Proposals and initiatives on this score abound: "reinventing government" and the many reengineering efforts around the world demonstrate how broadly recognized the problems are. Solving them may require a complete re-socialization of government, a radically modified view of government's role and its limitations. The situation calls for agility, but it is very difficult to institutionalize agility. Agility almost certainly requires individual discretion. And individual discretion is antithetical to formal rules and procedures.

Further complicating the challenge of contemporary governance is the public's increasing unwillingness to defer to the wisdom of government experts. Long gone is the era when Robert Moses could force rich landowners on Long Island to donate land along their property boundaries for a new parkway by threatening to build it through their front hallways if they refused, or could displace poor and minority communities to make way for his parkways.[24] Of course some projects hurt individual households, firms, and communities acutely. For that reason today it is much more difficult to build new projects. Affected parties have recourse to powerful legal instruments to challenge decisions, either on environmental or procedural grounds. And the existence of such recourse reflects public sentiment about the priority that should be given to new projects. As a result, the ability of government agencies to impose any view of what facilities should be built is sharply diminished.

THE NEED FOR A NEW APPROACH

Urban transportation planning in America is at a critical juncture. For more than half a century, transportation planners have relied on a suite of technical analytical models that are ill suited to the purposes for which they are used. The planners seek to address long-term trends in travel and the traffic and environmental impacts of such travel decades into the future—which reflects important and valid concerns. But the models' estimates are deeply flawed. They typically reflect heroic assumptions about adherence to land use plans, travel behavior, vehicle technology, and the fate of environmental toxics. Under the American system of government, no one has the authority to enforce such plans. As a result, planning has become as much a ritual as it is a scientific and objective enterprise.

These observations are not new to the transportation community. Most acknowledge the profound limitations of the modeling and analytical framework available for making planning decisions. Yet the models are part of a process that is institutionalized in funding programs and regulatory regimes. So deeply institutionalized is this framework that the author, at a recent

---

[23] Francis Fukuyama, Seminar, Department of Public and International Affairs, George Mason University, October 9, 1996.
[24] Caro, *The Power Broker*.

workshop on transportation planning research needs for the twenty-first century, confronted the view that it was more useful to consider restructuring government to improve transportation planning than it would be to consider wide-scale reform and reinvention of the transportation planning process. Consider the implications of that view: it is more useful to consider changing federal and state constitutions than it is to change the requirements of the Clean Air Act and Chapter 23 of the U.S. Code! Amend the constitution to bring it into accord with the requirements of transportation planning. Is transportation planning so important that American society should reorganize itself to make itself transportation-efficient?

The central theme of this book is that reform is needed—urgently needed! But reform should focus on rethinking the planning and decision-making process to fit the institutional and instrumental characteristics of the problem. Rather than reorganizing society to fit outmoded and archaic planning tools grounded in the best thinking of the 1950s, it is far better to reexamine first principles and consider reinventing urban transportation planning to fit the conditions, concerns, and best thinking of a new century. We are building the legacy systems of the future. We have a profound responsibility to strive towards providing future generations attractive, healthy, and safe communities. But we ought also to avoid saddling them with costly-to-maintain infrastructure that is ill suited to their needs and preferences.

To call for reform is not to denigrate efforts to improve existing tools and methods. Radical reform is a long-term enterprise, and much could be done in the meantime to improve predictive modeling and analytical techniques and their application. But that is not the objective of this book.

The challenge is profound because the central assumption that travel demand is predictable over a twenty-year planning horizon is no longer supportable. Prediction-based analytical planning has been recognized as sharply inadequate. Households and firms adapt their behavior to the range of possibilities available to them in maddeningly unpredictable ways. They steadfastly refuse to behave as planners anticipate they will. Just as battle strategy must be adapted to guerrilla techniques, so must transportation planners adapt to the unpredictability and uncontrollability of their constituent households and firms.

Some have argued that government should move beyond its traditional responsibility for the supply of infrastructure and begin to manage more actively the demand for facilities through pricing and regulation so as to induce households and firms to behave as planners want them to. "Demand management" techniques include a number of transportation control measures (TCMs), including car and van pooling, restricted free parking, improved transit, and restrictive land development, to favor alternatives to the single occupant vehicle (SOV).

One need only look as far as the American air traffic control system to see demand management in action—albeit on the verge of collapse. Federally employed air traffic controllers have movement-by-movement command authority over virtually every commercial passenger and air freight flight. No aircraft can push back from the gate without the approval of a federally employed controller.

While demand management surely has its place in particular situations, relying on it as the centerpiece of an urban transportation infrastructure development strategy raises a serious concern. Centering a policy on reducing the growth of a particular kind of travel—that in single occupant vehicles—seems questionable. What SOV travel is "bad"? What is acceptable? Who should decide? What kind of monitoring and enforcement would be required? Another example: Is free employee parking a subsidy to auto travel? Or is it, like air-conditioning, part of the working conditions of an office or plant? And, again, who should decide? How should a policy be monitored and enforced?

The sources of demand for travel, SOV and otherwise, are deeply embedded in the private firms and households that, together with government, produce, consume, and distribute goods and services in the economy. Greater government intervention in these private sources of demand seems a questionable strategy. Moreover, it is doubtful that the public will accede to government explicitly apportioning access to transportation infrastructure—especially roads, and especially when these same government institutions have begun to fall short in their management of the supply of infrastructure.

2

ORDER, EFFICIENCY, AND THE STRUGGLE AGAINST CHAOS

American transportation planning is deeply rooted in beliefs about beauty, order, and efficiency that were dominant during the first half of the twentieth century. Part of the argument of this book is that such beliefs are no longer appropriate organizing principles for transportation planning—indeed, they may never have been very appropriate—and that a new approach is badly needed.

This chapter examines the concepts of beauty, order, and efficiency that developed during the nineteenth century and became dominant during the twentieth century when the modern American highway program developed. Their origins date at least to the Italian Renaissance. At the turn of the twentieth century, they were apparent in the American Renaissance, the Progressive reform movement, and in the emergence of modern city planning.

ORDER AND EFFICIENCY IN THE ITALIAN RENAISSANCE

Urban infrastructure planners strive to contribute to the realization of beauty, order, and efficiency in an environment that is subject to the forces of disorder and chaos. Underlying their efforts are deeply held beliefs about the predictability and orderliness of natural and social systems.

Philosophy, art, and literature have long asked, "Is there an underlying order?" The classical debate between the Platonic and Aristotelian worldviews centered on the nature of that order.

> Plato, profoundly influenced by mathematics, thought of the world in terms of ideal and abstract perfect forms that lay behind the sloppiness of sensory experience. Aristotle, the biologist, elbow deep in dissections, thought of the world in terms of experimental categories and inductive generalizations. At its extreme, Platonism finds reality by de-

riving the observed world from abstract ideas. Aristotelianism at its extreme finds reality by deriving the essences from the observed world.[1]

Human settlements with orderly thoroughfares and buildings are evident in ancient times. A well-developed concept of order and rationality are readily apparent in the ancient cities of Mesopotamia, Egypt, and China, and in the Athenian Acropolis of the fifth-century B.C.[2]

The belief about order that has so profoundly influenced contemporary infrastructure planning emerged in the Renaissance. A central notion of the Renaissance attitude towards man's relationship with the world is that he is the "measure and mean of all things." That is, the logic and meaning of the universe only make sense in their relation to man himself. According to these humanistic (i.e., human-centered) ideas, the purpose of nature is to serve man, through whom "the order of God's creation" can become apparent. "Without man as a rational observer, such order would remain unseen. The perception of nature and intellectual command over its rules place human beings in a position of potential mastery within the physical world, giving them the means to make nature serve their purposes—even if these purposes are ultimately dependent upon God's decrees."[3]

Modern notions of environmental stewardship and conservation have now tempered this view of man's relationship with his environment. But the Renaissance concept of man's mastery of nature is of interest in understanding the urban transportation domain because it has so strongly influenced basic beliefs about the importance of urban form and the role of transportation infrastructure in influencing it.

The principles of Renaissance town planning emerged in the late Middle Ages, evident in the design of small towns as early as 1299. The basic principle was that the layout and design of streets, squares, and buildings should follow a rational order. The most important buildings should be arrayed around the central square, with buildings of lesser importance located further from the center. The widest streets would be those at the center, with the width decreasing with the distance from the center.

These principles were laid down as such by the humanist architect, art theorist, and writer Alberti in *The Art of Building* in the mid-1400s. "The principal ornament to any city lies in the siting, layout, composition and arrangement of its roads, squares and individual works; each must be properly planned and distributed according to use, importance and convenience. For without order there can be nothing commodious, graceful and noble."[4]

Alberti's successors took these principles a step further and envisaged the entire range of design, from column capital to city plan, as "governed by a rigorously proportional geometry."

---

[1] Harold Morowitz, *Entropy and the Magic Flute* (New York: Oxford University Press, 1993), 152.
[2] Lewis Mumford, *The City in History* (New York: Harcourt, Brace & World, 1961), 159.
[3] Martin Kemp, "The Mean and Measure of All Things," in *Circa 1492: Art in the Age of Exploration*, ed. Jay A. Levenson (Washington, D.C.: National Gallery of Art, 1991), 95.
[4] Alberti, cited in Kemp, "The Mean and Measure of All Things, 99."

The town plan also had important symbolic value insofar as it used the shapes and forms of so-called Platonic solids, which were thought to comprise the "building blocks of the universe."[5]

Figure 2-1
View of an Ideal City

The Walters Art Museum, Baltimore

The Renaissance notion of a city plan was that it revealed the *pre-existing* underlying structure of the universe for man's benefit, appreciation, and understanding. This order was the product of an exogenous force, indeed the ultimate exogenous force, "the order of God's creation."

Two key elements of the Renaissance concept of planning are important from the standpoint of urban transportation planning. First is the assumption of the existence of an underlying, externally defined order. Order is not the product of an individual's subjective judgment and discretion. The order is already there, waiting to be revealed.

Second is that the planner's responsibility is to discover and reveal that order. The quality of a planner's work depends on the beauty of the order it reveals. The more beautiful and pleasing the order, the better the plan. The planner thus occupies a position of great stature and power. His work is to reveal the beauty and order of God's creation. Other concerns—like impacts and costs—are secondary concerns.

CHAOS AND THE SEARCH FOR ORDER

At the turn of the twentieth century, half a millennium after Alberti, American community life was in a period of profound structural transformation.[6] Traditional institutions and mechanisms

---
[5] Kemp, "The Mean and Measure of All Things," 99.
[6] This discussion relies heavily on Robert H. Wiebe, *The Search for Order: 1877-1920* (New York: Hill & Wang, 1967).

of governance and control were eroding. National interests, national systems, and national corporations were eclipsing local control and independence. For many, the forces of chaos—self-interest, corruption, and irrationality—were ascendant.

Traditional small town life underwent dramatic changes in the latter half of the nineteenth century as a result of the Industrial Revolution. The population became increasingly urban and increasingly employed in manufacturing rather than agriculture. At the same time, the emergence of the national railroad system dramatically lessened the spatial and temporal separation—and independence—of rural America from the centers of government and commerce. Cutthroat competition between the railroads often led to seemingly irrational pricing patterns for the movement of agricultural products to market. Time was synchronized in 1883. Free rural delivery of mail was instituted in 1896.

Industrial organization also underwent dramatic changes due to the emergence of the large corporation and a national system for allocating capital. Prior to the last half of the nineteenth century, precedents for large private commercial organizations were relatively scarce. The only large organizations were the military, the state, and the church. Institutions were being stretched and rethought to accommodate the scale and scope of the technological systems that they governed. "Many [businesses] fought hard to impose some order on their affairs, and their successes, by contrast to the confusion around them, often gave men still groping in a half-light reputations for genius."[7]

Along with these changes in community life and industrial organization came a transformation of the national government to meet the public policy challenges of overseeing and regulating emerging national-scale industries and businesses. The executive branch was traditionally the weaker branch; Congress was the locus of national policy making authority, yet it was locally oriented, not yet strongly attuned to the national policy-making role. Large corporations, on the other hand, had truly national interests.

The effect of these changes on small towns was a loss of control over traditional aspects of everyday life to larger scale political and commercial institutions that were seen as seen as remote and often corrupt. The decimation of the small town of Sheldahl, Iowa, due to the location of a new rail crossing up the line a mile and a half is illustrative. The fate of towns and cities could turn on the discretion of remote forces little influenced by the wishes and values of local communities.

The search for order a century ago was a response to concerns on the same order of magnitude as those about globalization at the turn of the twenty-first century. Henry Adams captured the spirit of the period when he wrote, "Chaos was the law of nature ... order was the dream of man." Nature was "a chaos of anarchic and purposeless forces." Man clothed it "with the illusions of his senses, ... the splendor of its light, and the infinity of its heavenly peace."[8]

---

[7] Wiebe, *The Search for Order*, 22.
[8] Henry Adams, *The Education of Henry Adams* (1918; reprint, New York: Random House, Modern Library, 1931), 289. See also William Cronon, "The Trouble with Wilderness; or, Getting Back to the Wrong Nature," in

The contrast with today's fascination with chaos theory is stark. Chaos was perceived as a force of evil. Goethe's famous devil figure, Mephistopheles, in *Faust* is the "son of chaos" and sings, "Chaos is my one desire," in one musical adaptation.[9] He is "the spirit of chaos and disorder in the natural world."[10]

The role of science and enlightenment was to discover the underlying order of the universe. Darwin's theory of natural selection sought to explain the orderly evolution of the earth's species.[11] The economic theory of general equilibrium (1877) posited the existence of an orderly, predictable, steady-state price and quantity for each product and service. If the economic system were left undisturbed, a determinate general equilibrium would emerge.[12]

While this book does not explore these concepts in detail, it is important to recognize that early twentieth-century beliefs and assumptions about the existence of underlying order, and the legitimacy and goodness of man's search for it, permeated the full spectrum of efforts to solve social and economic problems. That view powerfully influenced the way the engineers and planners of that era structured and organized the definition of problems, analyzed them, and made decisions about transportation infrastructure.

THE AMERICAN RENAISSANCE

The impetus for new and transformed institutions often favored the rich and powerful in America, who saw the dawn of the twentieth century as a springboard for American improvement at home and hegemony abroad. America entered into a period of ascendancy in world affairs. The American Renaissance, as it came to be called beginning in 1880, captured an exuberant spirit of American nationalism. "America became the culmination of history for an age that believed in progress." "It appropriated images and symbols of past civilizations and used them to create a magnificent American pageant."[13]

> At the turn of the century Americans could look back over three generations of progress unparalleled in history. The nation had advanced, in Jefferson's prophetic words, to "destinies beyond the reach of mortal eye." The continent was subdued, the frontier was gone, and already Americans were reaching out for new worlds to conquer. From a small struggling republic, menaced on all sides, the nation had advanced to the rank of a world power.[14]

Uncommon Ground: Toward Reinventing Nature, ed. William Cronon (New York: W.W. Norton & Co., 1995), 69-90; John Lukacs, "The Gotthard Walk," *New Yorker*, December 23, 1985, 67ff.
[9] Arrigo Boito, "Mephistopheles," 1868 (Melville, N.Y.: Belwin Mills, n.d.), 83, 89.
[10] Jeffrey Burton Russell, *Mephistopheles: The Devil in the Modern World* (Ithaca, N.Y.: Cornell University Press, 1986), 157.
[11] Charles R. Darwin, *On the Origin of the Species* (1859).
[12] Léon Walras, *Elements of Pure Economics; or, The Theory of Social Wealth* (1874-77; reprint, London: Allen and Unwin, 1954).
[13] Richard Guy Wilson, "The Great Civilization," in *The American Renaissance, 1876-1917* (Brooklyn, N.Y.: Brooklyn Institute of Arts and Sciences, Museum, 1979), 12.
[14] Samuel Eliot Morison, Henry Steele Commager, and William E. Leuchtenburg, *The Growth of the American Republic*, 1930, 7th ed. (New York: Oxford University Press, 1980), II:266.

One manifestation of American nationalism was imperialistic. The Spanish–American War of 1898 liberated Cuba and gave into American possession Puerto Rico, Guam, and the Philippines. Teddy Roosevelt exercised his "big stick" policy, and the U.S. supported the secession of Panama from Columbia and the construction of the canal (1903-20).

A nationalistic exuberance was also apparent in the world of art, architecture, and landscape architecture. American robber barons saw themselves as analogous to French and Italian merchant princes. Stanford White, John La Farge, and other important contemporary figures emulated Renaissance principles by working in several media—art, architecture, illustration, and sculpture.

Bernard Berenson, who would later become the preeminent world scholar on the Italian Renaissance, wrote in 1894, "We ourselves, because of our faith in science and the power of work, are instinctively in sympathy with the Renaissance.... [T]he spirit which animates us was anticipated by the spirit of the Renaissance, and more than anticipated. That spirit seems like the small rough model after which ours is being fashioned."[15] The Italian Renaissance as a "small rough model"? The exhilaration—the hubris—is clear.

THE PROGRESSIVE ERA

While the wealthy and powerful flattered themselves about outshining the Medici, the dislocations of economic restructuring also inspired the emergence, beginning in the 1880s, of a movement of "Progressive" reformers who believed in rationality and science as a remedy for the chaotic forces of nature and corruption.

The Progressives argued for the establishment of decision making in the executive branch of government under the auspices of technical and scientific rationality. Decisions should not follow the self-interests of the robber barons or the backroom machinations of politicians. Rather, decisions should follow from scientifically and technically objective criteria that are not subject to manipulation and distortion for proprietary purposes.

Progressive ideology rested on the notions of efficiency and rationality. In general terms, the Progressive sense of order is embodied in the rational decision model, with its embrace of a systematic process for defining goals, identifying evaluation criteria, generating alternatives, evaluating alternatives against those criteria, and selecting the optimal alternative. This "gospel of efficiency" emerged in a broad range of disciplines and areas of inquiry.[16]

---

[15] Wilson, "The Great Civilization," 12.
[16] Samuel P. Hays, *Conservation and the Gospel of Efficiency: The Progressive Conservation Movement, 1890-1920* (Cambridge, Mass.: Harvard University Press, 1959).

Scientific Management

One of the most prominent manifestations of the gospel of efficiency was in industrial organization and management. The influence of efficiency expert and scientific management advocate Frederick Taylor and his followers Frank and Lillian Gilbreth was reaching its peak in the early twentieth century.[17] This approach involved careful time-and-motion studies of workers in various production activities in order to identify the most efficient configuration of workspace and task. The emphasis was on how to incorporate the individual worker into a maximally efficient production system. Not surprisingly, "Taylorism" was later much criticized for its effect of deskilling workers and dehumanizing work.[18] Taylorism also called for the application of scientific expertise not only to resources, home life, farms and agriculture, business and work, but also to city, state, and federal government.[19]

Another advocate of scientific management and efficiency was the eccentric economist and social critic Thorsten Veblen, who argued that

> the entire industrial system of the country should be under the systematic control of "industrial experts, skilled technologists, who may be called 'production engineers.'" He believed the nation's industry to be a "system of interlocking mechanical processes...." [He] called for soviets, or governing committees, of experts to take the management of the nation's industrial system away from parasitic financiers and inexpert entrepreneurs who were wasting the resources and manpower of the country through their...greed for profits and their competitive instincts.[20]

Standardization

In product design and manufacturing, the Progressive spirit and scientific management placed great emphasis on standardization, which had extraordinarily broad appeal and application. "Standardization is the outstanding note of the industry of this century," wrote one enthusiast in 1928, "[t]he trend is definitely away from mere *opinion*, or even *expert judgment*, and toward those qualities or characteristics of a commodity, verifiable by *measurement*, which assure maximum utility."[21] A true standard could never be the result of a mere opinion, but must be the result of scientific analysis.[22]

---

[17] Frederick W. Taylor, *Principles of Scientific Management* (New York: Harper & Bros., 1911); L. E. M. Gilbreth, *The Psychology of Management: The Function of the Mind in Determining, Teaching and Installing Methods of Least Waste* (New York: Macmillan, 1914; reprint, Easton, Pa.: Hive, 1973).
[18] See, e.g., L. L. Bernard, *Social Control in Its Sociological Aspects* (New York: Macmillan, 1939), 408-50.
[19] Samuel Haber, *Efficiency and Uplift* (Chicago: University of Chicago Press, 1964), 110-16, cited in Thomas P. Hughes, *American Genesis: A Century of Invention and Technological Enthusiasm, 1870-1970* (New York: Viking, 1989), 246-47.
[20] Veblen's *The Engineers and the Price System* appeared in serial form in the influential journal *The Dial* in 1919 and as a book in 1921. Hughes, *American Genesis*, 247-48, quoting from Veblen's *The Engineers and the Price System* (1921; New York: Harcourt, Brace & World, 1963), 72, 74.
[21] Norman Follett Harriman, *Standards and Standardization* (New York: McGraw-Hill, 1928), esp. 207-19. Emphasis in original.
[22] Gilbreth, *The Psychology of Management*, 143.

Standardization is based on the concept of the interchangeability of parts among identical machines or tools without additional tooling or refining. It is evident as early as the fifteenth century, when Gutenberg developed moveable type.[23] While often attributed to Eli Whitney in the 1790s, its use in manufacturing originated in 1765 when French general Jean-Baptiste de Gribeauval developed a system of arms production with interchangeable parts. Thomas Jefferson became familiar with it and promoted it in the U.S., where the War Department issued contracts for arms with interchangeable parts to Whitney and another manufacturer in 1798. While Whitney promoted the idea of interchangeable parts, he never successfully manufactured any.[24] Increasingly during the nineteenth century the incorporation of interchangeability in the process of manufacturing took hold in the production of sewing machines, bicycles, carriages, and, eventually, automobiles.

The standardization movement took the notion of interchangeability and combined it with the notion of standards for weight and measure to allow interchangeability across organizational boundaries and as a means of describing the quality and character of products and materials. Product standardization and labeling of consumer products were seen to have large potential benefits to consumers.[25] Housing standards were thought to have great potential for reducing housing prices and facilitating the creation of a secondary market in home mortgages.[26]

Government's role varied depending on the type of standard. Government might set health and safety standards through regulation. But it could also facilitate industrial activity through the encouragement of cooperative professional associations that set standards through voluntary consensus rather than government command and control. Herbert Hoover included this facilitating role within his notion of an "associative state," an idea that was to have an important influence on the formulation of the federal–state partnership in the development of highways.[27]

The synthesis between scientific management and standardization was superbly realized in the mass production facilities for the Ford Model T during 1913 and 1914 at the Highland Park factory in Detroit. The Highland Park factory was designed with extraordinary attention to the selection and placement of machine tools in order to speed automobile production. The shop floor was laid out to optimize the flow of materials and processes so that machines and processing equipment were located exactly where they were needed for the production of a particular component. The subsequent development of the continuous flow assembly line by 1915 carried the notion of production efficiency to another level.[28]

---

[23] Lewis Mumford, *Technics and Civilization* (New York: Harcourt, Brace & Co., 1934), 90.
[24] David A. Hounshell, *From the American System to Mass Production, 1900-1932: The Development of Manufacturing Technology in the United States* (Baltimore: The Johns Hopkins University Press, 1984), 28-32.
[25] Jessie V. Coles, *Standardization of Consumers' Goods: An Aid to Consumer Buying* (New York: Ronald Press, 1932).
[26] National Institute of Building Sciences, *Federal Regulations Impacting Housing and Land Development, Recommendations for Change, Phase I*, 11 (1981).
[27] See Ellis W. Hawley, "Herbert Hoover, the Commerce Secretariat and the Vision of an 'Associative State,'" *Journal of American History* 61 (June 1974): 116-40. Cited in Bruce E. Seely, *Building the American Highway System: Engineers as Policy Makers* (Philadelphia: Temple University Press, 1987), 68.
[28] On Ford and more generally the American system of mass production, see Hounshell, *From the American System to Mass Production*, esp. chap. 6.

But the Model T also revealed some of the dangers of standardizing a process too soon or too permanently. The standardization of the Model T at Highland Park had allowed tremendous efficiencies of production and dramatic reductions in price during its production life from 1908 to 1927. But the Highland Park production line was so highly optimized for the production of the Model T that retooling for production of the Model A proved to be a massive and disruptive operation.[29]

By the early 1930s, some healthy skepticism about the benefits of standardization had begun to emerge. "Premature" standardization could cause problems in cases of rapidly developing technologies, and bureaucracies sometimes sought to legitimize themselves by promulgating standards. Yet even among skeptics, solving a problem "once and for all" with a standard had enormous appeal.

In ironic contrast to the view today, the performance standard was seen as the most primitive type of standard because it failed to specify how a particular problem might be solved. Today, performance standards are often preferred, since they allow a problem to be solved in multiple ways using the most resources and methods.[30]

The standardization movement thus captures the Progressive spirit of reform that favored rational, scientific solutions to problems. But it also reflects a powerful underlying belief in the ability to solve problems permanently, once and for all, in accordance with an objective, "right" way. This belief in once-and-for-all solutions is closely related to the Renaissance concept of order.

Public Administration

The gospel of efficiency and the Progressive spirit were also extremely influential in establishing in the U.S. the field of public administration, which sought to apply rational, scientific management to government. Public administration in the U.S. is rooted in the separation of politics from administration, of policy making from the "details" of policy administration. The role of Congress, thus, was to oversee the formulation of policy, while that of the executive branch was to oversee the administration of that policy. This was the essence of the first U.S. article published in the field, by Woodrow Wilson in 1887.[31]

---

[29] Hounshell, *From the American System to Mass Production*, chap. 6.
[30] John Gaillard, *Industrial Standardization: Its Principles and Applications* (New York: H.W. Wilson, 1934). See also Mumford, *Technics and Civilization*, 386.
[31] Woodrow Wilson, "The Study of Administration," *Political Science Quarterly* 2 (June 1887): 197-222. By some accounts, Wilson's article was not widely influential until 1941, when it was reprinted in the same journal. The principle of separation of politics from administration was more explicitly developed in Frank J. Goodnow, *Politics and Administration: A Study in Government* (New York: Russell & Russell, 1900). Paul Van Riper, "The American Administrative State: Wilson and the Founders—An Unorthodox View," *Public Administration Review* 43 (November/December 1983), cited in Jay M. Shafritz and Albert C. Hyde, *Classics of Public Administration*, 3rd ed. (Pacific Grove, Calif.: Brooks/Cole, 1992). See also Richard J. Stillman II, *Public Administration*, 4th ed. (Boston: Houghton Mifflin, 1988), 405.

This view was an accepted norm until the crisis in executive authority that accompanied Watergate, the Vietnam War, and the unraveling of many Great Society programs. The fundamental notion was that once Congress set policy, objective, professional public administrators would oversee its execution according to the principles of scientific management and technical efficiency.[32]

The spirit of Progressive reform is evident from the stream of events that occurred between 1887 and 1906: civil service reform (1883), the establishment of the Interstate Commerce Commission (1887), the publication of Wilson's "The Study of Administration" (1887), establishment of the National Municipal League to fight local government corruption (1894), the publication of Taylor's *Shop Management* (1903), the formation of the American Political Science Association (1903), and the establishment of the Bureau of Municipal Research to further the management movement in government (1906).[33]

After World War I, public administration grew with the significant increases in the scale and responsibility of government at all levels. The basic principles of "orthodox" public administration were that "true democracy and true efficiency are synonymous, or at least reconcilable," "that the work of government could be neatly divided into decision making and execution, and that administration was a science with discoverable principles."[34]

CITY PLANNING AND THE CITY BEAUTIFUL

Progressive ideology influenced a broad range of other social and economic policy. Science, rationality, objectivity, and efficiency were the cardinal rules. One example of their realization was in the planning and design of cities

City planning was born out of mid–nineteenth-century efforts to improve sanitation through the provision of clean water and effective treatment of sewerage—objectives highly consistent with Progressive themes.[35] The rapid growth of cities during the late nineteenth century occurred at a time when few cities had authority over the disposition of private land. With few exceptions, cities could only prevent landowners from building in areas where the city expected to locate streets by two means. They could condemn the property and purchase through eminent domain (an expensive proposition). Or they could use moral suasion and refuse to provide city services (water and sewer) to buildings built in expected street rights-of-way. As a result, urban expansion to accommodate population increases often occurred helter skelter. One observer of Boston a century ago called it "pandemonium city."[36]

---

[32] Stillman, *Public Administration*, 405-06.
[33] Shafritz and Hyde, *Classics of Public Administration*, 9.
[34] Shafritz and Hyde, *Classics of Public Administration*, 39. First quotation is Dwight Waldo, quoted in same.
[35] On the history of city planning, see Mel Scott, *American City Planning since 1890* (Berkeley and Los Angeles: University of California Press, 1969); Wilson, "The Great Civilization"; John William Reps, *The Making of Urban America: A History of City Planning in the United States* (Princeton, N.J.: Princeton University Press, 1965).
[36] Patrick Geddes, quoted in Scott, *American City Planning since 1890*, 45.

A complementary development that focused on the aesthetics of cities was the City Beautiful movement, which was very much an outgrowth of the spirit of the American Renaissance. The fourth centennial of Columbus's discovery of America gave rise to a competition among American cities for a great celebratory exposition. Chicago, the winner, hosted (one year late) the 1893 World Columbia Exposition. This is generally recognized as the opening event of the City Beautiful movement. City Beautiful, inspired by the Columbia Exposition and, later, the McMillan plan for Washington, D.C. (1902) and the Burnham plan for Chicago (1906-1909), unleashed a widespread enthusiasm for the beautification of cities with classical buildings and parks. The objective of City Beautiful, in the words of one of its chief exponents, Daniel Burnham, was "to bring order out of chaos."[37] That is, it would bring the American city up to the standard commanded by the American Renaissance.

Thus, there was an overlap of purpose between the exponents of the City Beautiful and the city-planning Progressives. Both sought to respond to the chaos and disorder of the late nineteenth-century city. The difference was that Progressives responded with reform and professionalization while the City Beautiful movement responded with classical buildings and vistas.

Critics of the City Beautiful movement argued that it emphasized ornamentation and art at a time most city streets were not being cleaned and smoke ordinances were being routinely violated. Moreover, its focus on aesthetics provided virtually no solace to the poor or disadvantaged in American cities, a subject of increasing emphasis in the period. City Beautiful also raised concerns that it misappropriated resources, giving the public expensive buildings and plazas "whether they could afford them or not." Furthermore, while the City Beautiful movement led to the development of magnificent plans, their implementation was rare.

***

At the turn of the twentieth century—the "American Century"—order, efficiency, science, and rationality together comprised a powerful ideology. This ideology of order was a lens through which those concerned with roads and the development of highways saw their mission. It was a Progressive mission, informed by science and rationality and insulated from politics. The development of the highway program, and the process for analyzing and making decisions, is the subject of the next chapter.

---

[37] Wilson, "The Great Civilization," 21.

3

THE AMERICAN HIGHWAY PROGRAM TO 1956

The American response to the chaotic environment of the late nineteenth century was to develop institutions and intellectual frameworks to impose order. The notion of order is deeply embedded in the cultural identity of the period. In much of the domain of urban affairs and urban infrastructure, this imposition of order took the form of planning. Daniel Burnham's famous dictum, "Make no little plans; they have no magic to stir men's blood," goes on to say, "Let your watchword be order and your beacon beauty."[1]

One of the many expressions of this belief was in the enterprise of road and highway development. According to this view roads and highways should be developed on the basis of objective principles derived from scientific research and inquiry. The use of political criteria in locating and designing facilities was to be resisted energetically.

The American highway program was a forum for a series of profound conflicts between technical expertise and affected stakeholders. In the early years of the program, "scientific" road building distanced politicians from specific decisions about project location and design, and left them in the hands of engineers. This was fine as long as the engineers built roads that were acceptable to politicians and their constituents. But then the engineering diverged from what was acceptable politically, and political forces reasserted themselves, both through direct stakeholder participation, and through the intervention of elected politicians.[2]

---

[1] "Apparently this is a rephrasing by Willis Polk, an associate." Richard Guy Wilson, "The Great Civilization," in *The American Renaissance, 1876-1917* (Brooklyn, N.Y.: Brooklyn Institute of Arts and Sciences, Museum, 1979), 80.

[2] The history of the development of the U.S. highway program has received considerable study and is well documented in some excellent histories. See, e.g., Bruce E. Seely, *Building the American Highway System: Engineers as Policy Makers* (Philadelphia: Temple University Press, 1987); Mark H. Rose, *Interstate Express Highway Politics, 1941-1989*, rev. ed. (Knoxville: University of Tennessee Press, 1990); American Association of State Highway and Transportation Officials, *The States and the Interstates: Research on the Planning, Design and Construction of the Interstate and Defense Highway System* (Washington, D.C.: The American Association of State Highway and Transportation Officials, 1991). There is also an official history, U.S. Federal Highway Administration, *America's Highways, 1776-1976: A History of the Federal-Aid Program* (Washington, D.C., 1977). For a history of nineteenth-century municipal street development, see Clay McShane, *Down the Asphalt Path: The*

THE FIRST-GENERATION HIGHWAY PROGRAM

Until 1916, the federal government exercised almost no control over road and highway activities. Authority for road construction and maintenance resided in the individual states or in their subsidiary counties and municipalities. According to the reform rhetoric of the period, county road development tended to be piecemeal, with improvements often following a political calculus. Funds were often expended over much of the road mileage within a county, maximizing its visibility and political value without much regard for lasting improvements. Alternatively, improvements occurred on roads of political importance, that is, roads serving politicians or important constituents.

Some state highway departments had jurisdiction over their main rural roads. Many, however, lacked authority or funds. Some state constitutions prohibited the state from making any internal improvements whatsoever, delegating that responsibility completely to localities. Overall, authority for roads was fragmented across small jurisdictions, and development tended to be insubstantial or piecemeal.

The federal program began with the creation of the Office of Road Inquiry in the U.S. Department of Agriculture in 1893, the year of the World Columbia Exposition in Chicago. Its purpose was to promote the development of good roads in rural areas. In its early years, the program had virtually no money to give out for actual road building. Instead, it promoted the benefits of "scientific road building" to rural communities.

One important promotional technique was the "object lesson road," a short segment of improved road built to demonstrate its value to a local community. The "scientific" features of these roads, which were usually unpaved, were proper drainage of the road surface, proper preparation of the roadbed, and proper location, all of which served to improve the longevity of improvements and the usability of the road in bad weather.

Over the next twenty years, the focus of the program shifted from pure promotion of good roads to direct federal support for construction. In the first attempt at direct support, Congress allocated funds in 1913 to support the construction of roads in cooperation with counties. The enterprise foundered badly. The legal powers of state and local officials varied widely among the states, and negotiating the terms and conditions of federal support with each county proved extremely burdensome.

Automobile and the American City, Columbia History of Urban Life (New York: Columbia University Press, 1994). This author and others have written on the evolution of the planning approaches and methodologies in the highway field. See Jonathan L. Gifford, "The Innovation of the Interstate Highway System," *Transportation Research A: Policy and Practice* 18A (1984): 319-32; Jonathan L. Gifford, "An Analysis of the Federal Role in the Planning, Design and Deployment of Rural Roads, Toll Roads and Urban Freeways," (Ph.D. diss., School of Engineering, University of California, Berkeley, 1983); Edward Weiner, *Urban Transportation Planning in the United States: An Historical Overview*, <http://www.bts.gov/tmip/papers/history/utp/toc.htm>, 5th ed. (Washington, D.C.: U.S. Department of Transportation, September 1997).

The second attempt was much more successful. In 1916, Congress created the Federal-Aid Highway Program, which is still the basic program for federal support of highways.[3] The Federal Aid Road Act of 1916, which first established the program, identified the states as recipients of federal funds, subject to various conditions. Reliance on the states as the senior partners in the federal highway program continues to the present, although local officials have played a stronger role since the early 1960s. The Intermodal Surface Transportation Efficiency Act of 1991 (ISTEA) attempted to strengthen the role of metropolitan regional planning organizations in decision making, and its successor, the Transportation Equity Act for the 21st Century (TEA-21), extended that effort.[4]

The 1916 act addressed federal concern over jurisdictional fragmentation by encouraging the creation of a statewide agency that could bridge the jurisdictional fragmentation among the county governments. An important feature of these state highway departments was that they were technically oriented, staffed by engineers, and managed in accordance with the principles of scientific management and administration. The law required states to create a state highway department to serve as a partner in the federal program that would have sufficient authority to supervise the expenditure of the funds. They could not simply pass federal funds to localities. It also required them to match federal funds dollar for dollar, which gave them a vested interest in each project. Motivated by these two provisions, all states had created and provided funding to their highway departments by 1917.

World War I highlighted the importance of roads to the nation's defense capabilities. It brought shortages and dramatic price increases for road building materials, which curtailed road building. At the same time, poor performance by the railroad system shifted a heavy burden of war materiel onto the infant road system.

The 1920s

The road system's strategic importance, and the need to repair war-damaged roads, strengthened the constituency for federal support of road construction. In 1921 Congress enacted its second major federal road authorization. In addition to providing funds, the Federal Highway Act of 1921 addressed federal concerns about piecemeal highway development by requiring state highway departments to designate a statewide system of highways that could not exceed 7 percent of its total rural road mileage. This system required federal approval, and federal funds were available for improvements only on this system. Further, expansion of the "seven-percent system" was only permitted when the state had improved the entire length of the original system.[5]

---

[3] *Federal Aid Road Act of 1916*, 39 *Stat.* 355 (11 July 1916).
[4] *Intermodal Surface Transportation Efficiency Act of 1991*, P.L. 102-240, 105 *Stat.* 1958 (18 December 1991). *Transportation Equity Act for the 21st Century* (P.L. 105-178, 112 *Stat.* 107-509, 9 June 1998) authorizes the federal surface transportation programs for highways, highway safety, and transit for the six-year period 1998-2003. *TEA 21 Restoration Act* (Title IX, P.L. 105-206, 112 *Stat.* 834-68, 22 July 1998) provided technical corrections to the original law. This book refers to the combined effect of these two laws as TEA-21.
[5] *Federal Highway Act*, P.L. 67-87, 42 *Stat.* 212 (9 November 1921).

Allowing states to designate the system, with federal approval, preserved some state autonomy. Limiting federal expenditures to this designated system prevented the dispersion of federal funds over too great a mileage. And requiring the improvement of the entire seven-percent system prior to system expansion provided a strong deterrent to selective neglect of regions lacking political clout.

One of the major controversies in the debate over the 1921 act was the extent to which the system would provide access from rural areas to railheads and cities, as opposed to connecting cities with each other. Rural interests generally favored "getting the farmer out of the mud," while urban and commercial interests and, interestingly enough, bicyclists, favored improvements between cities and from cities to their hinterlands. The resulting compromise was that the Federal-Aid Highway System would consist of a "primary system" of highways that connected the principal cities within a state and across state boundaries, and a "secondary system" that would serve rural areas. Together they could not exceed 7 percent of the state's overall road mileage, with the primary and secondary systems limited to 3 and 4 percent, respectively.

Highway system development flourished after the war. States not only complied with the provisions of the federal program, they exceeded them. By 1934, 94 percent of the states' seven-percent systems had received some measure of improvement, almost 40 percent of the system without the benefit of federal assistance.[6] States concentrated their expenditures on limited, designated systems, and responsibility for rural highways shifted increasingly to the state highway departments. State highway budgets increased more than three-fold between 1921 and 1930, while federal and local highway budgets remained relatively constant. And in expending their highway funds, the states concentrated their funds on their limited systems, rather than merely increasing the mileage of state-administered roads to keep pace with budget increases.[7]

The state gasoline tax was an important impetus in the development of state highway programs. With the increasing ownership and use of the automobile during the 1920s, it provided a constantly increasing revenue source for highway system development. The gas tax appeared first in 1919, and spread to thirty-three states by 1923, and all states by 1929. Most states dedicated gas tax revenues to highways. Gas tax revenues, in combination with state vehicle registration fees, contributed 80 percent of the increase in state highway departments' budgets between 1921 and 1929.[8]

One of the most attractive aspects of the gas tax was its immunity to economic cycles. Between 1925 and 1950, through the Depression and the gas and automobile rationing of World War II, state revenues from user imposts declined only twice, for two years in the early Depression and for two years during World War II.[9] The linkage between state highway expenditures and state user imposts thus insulated highway budgets through the most severe economic dislocations of the first half of the century.

---

[6] U.S. Federal Highway Administration, *America's Highways*, 126.
[7] Gifford, "An Analysis of the Federal Role," 50, 53.
[8] Gifford, "An Analysis of the Federal Role," 59-60.
[9] Gifford, "An Analysis of the Federal Role," 97.

The 1920s, then, witnessed a dramatic rise in the ascendancy of the state highway departments, grounded in part in the provisions of the federal-aid program, and fueled by the use of earmarked user imposts. It also witnessed a dramatic improvement in the road system, to some extent guided by the programmatic requirement of the seven-percent system.

The Early Depression: 1930-1936

The Great Depression that struck in the early 1930s severely tested the Progressive principles embodied in the Federal-Aid Highway Program. Unemployment moved to the top of the federal agenda, and the federal-aid program did not measure up very well as an employment program. States spent a relatively small portion of their federal-aid funds on labor, with a large share for planning and engineering studies and road building materials and equipment. The state's rate of expenditure of authorized federal-aid funds was also often rather slow, since the program emphasized careful, systematic engineering and construction, rather than quick expenditures.

The Bureau of Public Roads did the best it could to sell the program as an employment program, while preserving its procedural and technical integrity. It provided estimates of direct and indirect employment, and defended the use of specialized road building equipment in lieu of hand labor. Meanwhile, Congress attempted to improve its effectiveness as an economic stimulus. One major change was allowing states to use federal-aid funds for rural secondary roads and urban extensions of the rural seven-percent system, which greatly expanded the length of roads eligible for federal support.[10]

But Congress also authorized road spending under a variety of other public works programs, most significantly in the Works Progress Administration (WPA, later the Works Projects Administration). The WPA's road program was not bound by the restrictions of the federal-aid program. The WPA could work directly with counties and municipalities, while the federal-aid program worked solely with the state highway departments. Also, the WPA program was not constrained in the roads to which it lent support, while the federal-aid program was limited to the designated highway systems, although those systems had increased significantly in length. WPA funds were largely restricted to wages, however, and could not be used to purchase equipment or materials.

But the federal-aid program survived these challenges intact, preserving the senior status of state highway departments as partners in road building, and the principle of a limited, interconnected system of highways.

---

[10] Gifford, "An Analysis of the Federal Role," 85-104.

The Late Depression: 1937-1941

Beginning in 1937, events took a sharply different turn. The early Depression saw revisions to the federal-aid program and parallel public works initiatives in the WPA. The two programs complemented each other. But during the latter years of the Depression, the two programs were pitted against each other in direct competition for funds.

The conflict was precipitated by President Franklin Roosevelt's desire to balance his budget. To avoid raising taxes, he needed to cut expenditures, and he chose to try to cut the federal-aid program rather than the WPA budget. The basis of his choice, as he stated it, was that the WPA programs employed more of the unemployed than did the federal-aid program, and that WPA therefore had higher priority in his budget.[11] It may also have been significant that the WPA was a product of the Roosevelt administration, whereas the federal-aid program was a carry-over from the past that had sought to preserve its independence from political control.

While Roosevelt's attempt to cut the federal-aid program was ultimately unsuccessful, it heightened the political stakes for the program from autonomy to survival. This shift in the political stakes precipitated a marked change in the Bureau's behavior, which shifted from defending the program as it stood to suggesting a bold, new initiative, which eventually became the Interstate highway system.

The shrinking budgets of the Depression had forced the Bureau to identify and protect those aspects of the federal-aid program that were most important. First among these was the state–federal partnership, which proscribed direct relations between the Bureau and local governments. Planning and decision making were the province of engineers, technical experts who were above politics. By the early 1940s, a volume celebrating the thirtieth anniversary of the American Association of State Highway Officials (AASHO) referred to the "Secular Trinity" of AASHO, the state highway departments, and the Bureau of Public Roads.[12]

A second central plank of the federal-aid program was the concentration of funds on a limited, designated, and interconnected system of highways. The Depression had effectively relaxed this provision by expanding the number of systems eligible for federal aid to include a rural secondary system and a system of urban extensions. But the underlying principle still remained. Highways had to be designed as a system, and that meant limiting the length and number of highways that made up that system.

These two bedrock principles were to have a profound effect on the subsequent development of American highways. Both principles tended to exclude local considerations, the first explicitly and the second effectively, if more subtly.

---

[11] U.S. Bureau of Public Roads, *Toll Roads and Free Roads: Message from the President of the United States...*, H. Doc. 272, 76th Cong., 1st sess. (1939).

[12] The phrase is used in Samuel C. Hadden, President, AASHO, "Introduction," in American Association of State Highway Officials, *The History and Accomplishment of Twenty-Five Years of Federal Aid for Highways: An Examination of Policies from State and National Viewpoints*, Presented before the 30th Annual Convention (Cincinnati, Ohio, November 28, 1944), 4.

The limited system principle excluded local considerations because the definition and design of the system took place at the state and national level. Cities and towns were nodes on a state network. What took place within those nodes was effectively outside the system. This tended to privilege traffic at the spatial resolution of the network, that is, at the state and national level. Local traffic was outside the federal-aid system. And local officials were outside the Secular Trinity.

According to these principles, the Bureau's bold new initiative to fend off White House efforts to reduce or eliminate its budget would necessarily be constructed under the supervision of the state highway departments and according to a limited-system network configuration. However valuable these principles may have been in the first two decades of the federal-aid program, they were to have important if not devastating impacts later.

PLANNING THE INTERSTATE SYSTEM

Many associate the Interstate Highway system with the Eisenhower administration and the creation of the Highway Trust Fund in 1956. Eisenhower's support was in fact critical in securing passage of the funding package for the Interstate in 1956. And his recognition of the importance of good highways dated back as far as 1919, when he traveled in a 62-day military road convoy between Washington, D.C., and San Francisco. In World War II, he experienced the value of a more advanced highway system as the Allies advanced into Germany.[13]

But the conceptualization of the Interstate is really a product of the Roosevelt administration, and its design, that of the Truman administration. The major studies and policy debates began in 1937; Congress established the system in law in 1944; and the designation of its routes and specification of its design standards were complete by 1947.

The idea of a national system of strategically important highways dates to the 1920s. Based on the importance of highways in World War I, General John Pershing, in cooperation with the Bureau, developed a map of important military highways in 1922.[14]

By 1937, the Bureau of Public Roads was actively investigating a new national system of expressways. Interest in such a system derived from several quarters. President Roosevelt requested in early 1937 that the Bureau prepare a feasibility study of a 14,000-mile national system of toll roads, three running east–west, and three north–south.[15] Also in 1937, the House Committee on Roads held hearings on the same proposal and requested, in its next biennial au-

---
[13] Richard F. Weingroff, "Broader Ribbons across the Land," special issue of *Public Roads* (June 1996): 12–13; John Steele Gordon, "Through Darkest America," *American Heritage* (July/August 1993): 18ff.
[14] U.S. Federal Highway Administration, *America's Highways*, 142–43.
[15] U.S. Federal Highway Administration, *America's Highways*, 136.

thorization bill, a Bureau study of such a system.[16] By then, however, the Bureau had already submitted a draft of the report to the President for approval.[17]

The Bureau's analysis of the toll proposal, and its recommendation of a toll-free system that became today's Interstate, has been widely hailed as an engineering *tour de force*. It made unprecedented use of nationwide data on traffic flows, origins and destinations, trip purposes, and vehicle types that the Bureau had collected in highway planning surveys during the 1930s, supplemented by other surveys sponsored by the WPA.

These data were thought to provide an objective analytical basis for evaluating the viability of a national toll network, which it found to be infeasible, and a toll-free system, which it found to be feasible. The highways proposed in both cases were of the highest engineering design, with limited access and gentle grades and curves that could safely accommodate high-speed cars and trucks.

While presented as the epitome of scientific objectivity, the Bureau's report, published as *Toll Roads and Free Roads,* was also a vehicle for some potent political and ideological positions. It asserted that the proper way to build an American superhighway system was with public institutions, the same Secular Trinity that had built the first-generation highway system. Toll roads under the control of the private sector or quasi-independent authorities were not only financially infeasible, but also "a return to nineteenth century thinking." This position carried the day, and tolls played an important but minor role in the development of the second-generation road system.

Toll Roads and Free Roads also asserted forcefully that the nation's biggest highway problems were in its cities, and that the next generation of highways had to respond to urban travel demand, not just interregional travel. This position did not carry the day, with fateful results in many American cities. FDR opposed a major urban highway initiative because he saw it as a giveaway to the cities. Congress—dominated at that time by rural interests—was not particularly open to the idea either. Even the state highway departments, the Bureau's stalwart partners in the Secular Trinity, were not quick to come forward with support.

Of course, the Interstate has ended up serving a huge share of urban highway traffic. But the design of the Interstate facilities was so ill suited to many older cities that it precipitated a furious backlash against the urban Interstates, and against urban highways in general. But that gets ahead of the story.

---

[16] U.S. Congress, *Roads,* hearings on H.R. 7079, H.J. Res. 204, and H.J. Res. 227, 75th Congress, 1st sess. (1937).
[17] Thomas H. MacDonald, Letter to Colonel James Roosevelt, with attachments, 16 April 1938, official file 129, Franklin D. Roosevelt Library, Hyde Park, New York.

Toll Roads versus Free Roads

One of the most significant conclusions of *Toll Roads and Free Roads* was that a toll-financed system was financially infeasible. The Bureau used the data from its traffic surveys to estimate the return on the construction of the proposed three-by-three toll road network, which it found to be only about 40 percent of its cost. But this calculation relied on some important assumptions that turned out to be wrong. The toll system, had it been built, would likely have returned a profit.[18]

The two critical assumptions in the Bureau's analysis were the rate of traffic growth and the willingness of highway users to pay tolls. The mistaken assumptions were likely a result of several factors. Honest ignorance was part of it. Virtually no one in the late 1930s anticipated the phenomenal growth in highway usage that eventually occurred.

Nor was the Bureau alone in its skepticism about the public's willingness to pay tolls. The Pennsylvania Turnpike, which opened in 1940, had considerable difficulty selling its construction bonds to a skeptical market inexperienced with toll road financing. Indeed, FDR's federal Reconstruction Finance Corporation (RFC) bought some of the bonds to allow the project to go forward.

Another part of the Bureau's argument against toll roads was that net benefits would be greater without tolls because of the design constraints and costs associated with toll collection. In areas with relatively sparse traffic, the construction and especially the operating costs of tollbooths dictated a minimum economic interchange spacing on the order of twenty to thirty miles. Such large interchange spacing would generally exclude short trips because of the need to travel to the nearest interchange. To serve these shorter trips, the Bureau argued, it was necessary to construct parallel non-expressway roads, an additional cost imposed on the toll alternative. A toll-free system, on the other hand, could have more closely spaced interchanges, on the order of ten miles, avoiding the cost of constructing "duplicative" facilities.

This objection derived from a well-accepted principle of cost–benefit analysis that when the marginal costs of serving another user are low, as they are in a rural setting with sparse traffic and no toll collection, the benefits gained by users can exceed the costs of providing the service.[19] If gas taxes or general revenues could finance a toll-free system, then it could generate many benefits from short trips that could not be claimed by a toll alternative.

But bureaucratic self-interest, in combination with strongly held beliefs about the proper role of government in the provision of highways, may also have played an important part in the Bureau's position. Evidence contradicting the Bureau's assumptions about willingness to pay tolls soon began to accumulate. The Pennsylvania Turnpike quickly experienced traffic levels much higher than anticipated. The RFC sold its bonds at a profit once the bond market had an oppor-

---
[18] Gifford, "An Analysis of the Federal Role," 114-30.
[19] Duncan MacRae, Jr. and James A. Wilde, "Perfect Markets, Imperfect Markets, and Policy Corrections," in *Policy Analysis for Public Decision* (1979; reprint, Lanham, Md.: University Press of America, 1985), chap. 5.

tunity to see the public's enthusiasm. By 1941, the evidence was fairly clear that the public was enthusiastically willing to pay tolls.

But that evidence was never incorporated into the planning of the Interstate. The Bureau dismissed the Pennsylvania Turnpike as an aberration and held firmly to its position that toll roads would work only in very specialized locations with high traffic density, and that the benefits to the public would be much greater with a toll-free system. While a discussion of toll financing arose again during the Eisenhower administration prior to the passage of the 1956 act, the Bureau's position remained in place until the 1980s.

Why would the Bureau of Public Roads—the quintessential model of Progressive scientific objectivity—fail to reexamine this new evidence as it began to accumulate? There is no hard evidence, unfortunately. Part of it may have been naked bureaucratic self-interest. The Bureau and its engineers had no interest in remaining the custodians of the first-generation highway system while independent or private toll authorities built the next generation.

A more important reason may have been fundamental beliefs about the appropriate role of government and the integrity of the private sector. Roads, in their Progressive worldview, should not be built for private profit but rather for the public interest. Roads should not finance the private fortunes of a new generation of Robber Barons.

This ideological opposition to toll roads was partly rooted in the pricing practices of private franchise owners of toll facilities in the past. Thomas MacDonald, chief of the Bureau of Public Roads from 1919 to 1953, vehemently opposed "private profit" from road users.[20] Furthermore, nineteenth-century experience with toll roads in Great Britain had led Parliament to dissolve many road franchises in response to what were perceived as "confiscatory" prices, and in their place instituted national support for local roads.[21] Nineteenth-century U.S. experience with (publicly subsidized) private construction of the national railroad network had led to an expensive, redundant route structure and ruinous competition.[22] The Progressive belief was that public institutions, using objective engineering and expertise, should build a system that would bring the greatest benefits to all.

There was also the matter of "the system." While *Toll Roads and Free Roads* advocated a national highway program with a strong urban focus, the Bureau held strongly to the view that highways should be integrated into a national system. Operationally what that meant was that some public institution had to define a network of facilities, national in scope, that would best serve the public interest and provide a rational basis for facility investments. Such a system would not—could not, in their view—evolve out of the uncoordinated actions of private companies seeking a profit, or independent communities seeking to improve their local or regional highways. The system focus, scoped at a national level, was essential.

---

[20] Thomas MacDonald, "The Freedom of the Road," *America's Highways* 8, no. 1 (January 1929): 6, 7, quoted in U.S. Federal Highway Administration, *America's Highways*, 116.
[21] Charles L. Dearing, *American Highway Policy*, pub. no. 88 (Washington, D.C.: Brookings Institution, 1941), 22.
[22] David Haward Bain, *Empire Express: Building the First Transcontinental Railroad* (New York: Viking, 1999).

Urban versus Interregional Highways

A second major conclusion of *Toll Roads and Free Roads* was that the nation's most critical highway problems lay in urban areas, and that only modest improvements were needed for intercity highways. The balance between the rural, intercity portions of the Interstate and its urban portions turned out to be a source of considerable contention. The resolution of this issue had profound implications for both the urban and rural impacts of the Interstate program.

The non-toll alternative highway system put forth in *Toll Roads and Free Roads* was 27,000 miles in length, almost twice the length of the proposed toll system. This national system would require a large program of upgrading existing rural highways, and an even larger program of new superhighways in urban areas.

The Bureau's original draft proposal was discussed at a cabinet meeting on March 29, 1939. President Roosevelt asked for revisions to place less emphasis on urban highways and more emphasis on rural highways "... so as to make it less probable that the cities will be able to get from the Congress federal funds for doing work for which the cities should pay [and] to lay more emphasis on through highways as a mechanism for National Defense."[23] The Bureau hastily revised the proposal and resubmitted it on April 11. The revised proposal increased the mileage of new rural highways instead of supporting improvements to existing rural highways, as had the original proposal. But it left intact the proposal for new urban highways, and the president submitted it to Congress in that form in 1939.[24]

This was not in fact a real resolution of the conflict. The urban sections of the proposal remained intact, and the rural sections were added. But the revisions did not address Roosevelt's concern over a continuing urban focus. The Bureau continued to advocate a strong urban program throughout the 1940s and 1950s.

Contention over the issue continued with the next major study. In April 1941, Roosevelt created the National Interregional Highway Committee with the explicit charge to investigate and report on the feasibility of a national system of interregional highways, with no reference to the urban recommendations in the earlier study. But Bureau chief Thomas MacDonald chaired the committee. While giving lip service to the interregional limitation of the committee's charge, the committee's report, *Interregional Highways*, gives substantial attention to investigating local urban highway needs.[25] This report, submitted to Congress by the President in January, 1944, served as the basic support document for the creation of the Interstate system in the *Federal Aid Highway Act of 1944*.[26]

---

[23] Henry Wallace, Secretary of Agriculture, Dictation to Miss Batchelder, March 29, 1939, quoted in Mary Huss, Personal Secretary to Secretary Wallace, to Henry Kannee, March 29, 1939, Attachment 3, Maude Poulton, Memorandum to Mary Huss, Personal Secretary to Secretary of Agriculture Wallace, April 6, 1939, with attachments, Official File 129, Franklin D. Roosevelt Library, Hyde Park, New York.
[24] Gifford, "An Analysis of the Federal Role in the Planning, Design and Deployment of Rural Roads, Toll Roads and Urban Freeways," 131-34.
[25] U.S. House Committee on Roads, *Interregional Highways*, Report of the National Interregional Highway Committee, 78th Cong., 2nd sess., 12 January 1944, H. Doc. 379.
[26] "Federal Aid Highway Act of 1944," 58 *Stat.* 838 (20 December 1944).

The Bureau's position favoring urban expressways as part of the Interstate system was included by Congress in the 1944 act. But it was something of a pyrrhic victory. The Interstate would serve urban highway needs with two kinds of facilities, urban extensions of intercity routes and circumferential highways to distribute interregional traffic within an urban area without passing directly through the central business district.

But urban Interstates alone could not serve urban travel needs well. Both MacDonald and his deputy, H. E. Hilts, were acutely aware that the development of urban Interstates without adequate development of other urban transportation facilities would seriously compromise the value of those Interstates to urban areas. Both discussed this danger in journal articles where they admonished the cities and the states to make the complementary improvements in other facilities—other highways, surface streets, parking facilities, and transit—that were necessary for a well-balanced urban transportation system. The construction of the interregional routes in urban areas without the necessary complementary facilities, they argued, would overconcentrate traffic on those routes, leading to the rapid development of congestion and a concomitant deterioration of performance.[27] How right they were!

Design Standards

One of the key determinants of the impact of the urban Interstates was their design standards. Both *Toll Roads and Free Roads* in 1939 and *Interregional Highways* in 1944 laid great emphasis on adherence to the highest standards of engineering design. The most important design criteria were design speed, the limitation of access, grade-separated intersections, and capacity determination.

Design speed governs the vertical and horizontal curvature of a highway's alignment. For high-speed driving to be safe, curves must be gradual, both to maintain traction and to allow an adequate time for braking if an obstruction is hidden by a curve. The higher the speed, the more gradual the necessary curvature. As travel speeds had increased during the 1920s and 1930s, the design deficiencies of many of the first-generation highways had become painfully evident in accident and fatality statistics.

If trucks were anticipated to use the highway, safe design usually indicated even more gradual curves. While truck drivers typically sit higher off the ground and can therefore sight an obstruction in the road over a vertical curve from a greater distance than a car, they require a greater distance to come to a stop due to their weight. Moreover, their traction characteristics would require more gradual horizontal curves at a given speed than a car. Safe design also had to accommodate travel in adverse weather, when roads were slippery. Ideally, then, safe highway design would accommodate large trucks traveling at high speeds on wet pavement.

---

[27] H. E. Hilts, "Planning the Interregional Highway System," *Public Roads* 22 (June 1941): 85; Thomas H. MacDonald, "Flatten Out Those Traffic Peaks: Highways of Tomorrow Will Avoid Big City Congestion and Provide Safe Speed in the Country," *Motor* 76, no. 4 (October 1941): 102, 248-50.

The desire for limited access grew out of concern about the effects of roadside activities on a highway's throughput. As roadside development increased, vehicles entering and exiting the traveled way tended to reduce the throughput of the road, leading to congestion in heavy traffic. Limited access avoided interference of roadside activities with through travel. Access to the traveled way was allowed only at controlled intersections or grade-separated interchanges—that is, at points subject to careful engineering design rather than simply curb cuts into adjacent parking lots and businesses.

While fairly simple conceptually, access control raises a complicated legal issue. Common law gives a presumption of access to any public road by an adjacent landowner. So restricting access requires either compensating a landowner for that loss or providing an alternative, either through an access road that runs alongside the controlled access facility, or through another facility. Generally speaking, taking away access from an existing roadside business imposed a much greater hardship—and therefore payment—than not allowing access to a new facility by an adjacent landowner. As a result, limited access roads were often less costly on new rights-of-way.

Neither limited access nor speed-based geometric design was new—both were nineteenth-century innovations implemented for carriageways in Central Park in 1876, among other places.[28] In 1912, Senator Coleman du Pont of Delaware had constructed, at his own expense, a 100-mile "boulevard" with a 200-foot-wide right-of-way that bypassed towns and had grades and alignments consistent with those used in expressway construction in 1939, more than twenty-five years later. Ironically, the state of Delaware resisted his efforts to present the highway as a gift to the state.[29] The Bronx Parkway, designed in 1917, had controlled access and, when opened to traffic in 1923, was extremely popular.[30] While these highways stand out as prescient examples of highway design, the value of their design features on motorways was largely unappreciated until the 1930s.

An important philosophical underpinning of limited access is the privilege it affords to through traffic over local access and egress. The underlying logic is that the higher purpose of the highway is to move traffic, not to provide access to property. Access to property is subordinated to "lower order" collector streets.

Such privileging of through traffic is also evident in another of the Interstate's design standards, the use of grade-separated intersections. Grade-separated intersections such as the cloverleaf were a design feature that allowed local traffic to enter and exit the traffic stream with only minimal disruption to through traffic, at least under uncongested conditions. But this design gave the right-of-way to through traffic and required entering (local) traffic to yield.

---

[28] McShane, *Down the Asphalt Path*, 33 passim.
[29] W. W. Mack, "Annual Convention Address by the President of the Association," *American Highways*, vol. 18, no. 4 (October 1939): 4.
[30] U.S. Federal Highway Administration, *America's Highways*, 132-33.

Alternative intersection designs strike a different balance between local and through traffic. Signalized intersections provide great flexibility in the relative priority given to various traffic movements through the length of the signal cycle, the particular traffic movements included in the cycle, and the duration of each "split." Roundabouts and traffic circles afford all traffic roughly equal priority through the intersection, since all traffic approaching the circle must generally yield to traffic already in the circle, whether local or through.

The space requirements for a grade-separated interchange also depended on the design speed of the highway. Grade-separated interchanges generally featured acceleration lanes so that entering traffic could reach cruising speed before merging with through traffic, and deceleration lanes so that exiting traffic could begin to decelerate after leaving through traffic. The higher the design speed, the greater the required length of the acceleration and deceleration lanes. The design speed also affected the minimum interchange spacing, since it was generally desirable to have some minimal distance between the end of the acceleration lane from one interchange and the beginning of the deceleration lane from the next interchange.

The number of lanes on a highway was also an important determinant of performance. *Toll Roads and Free Roads* and *Interregional Highways* had concluded—with seemingly sound logic—that the source of congestion on many urban highways was too few lanes, that is, inadequate capacity. To avoid this eventuality in the future, Interstates would be designed to have adequate capacity for twenty years, roughly the lifetime of the pavement material. There is also an almost poignant tone of self-castigation in *Toll Roads and Free Roads* that some of the highways built during the first-generation program had remaining physical life but had become functionally obsolete due to either congestion or increased driving speeds.

It is striking that that the respect for engineering effort and engineering materials was so great as to motivate the heroic assumption that traffic could readily be forecast for twenty years using objective scientific techniques. In urban areas, where traffic was increasing at dramatic rates, forecasts of rapid growth compounded over twenty years indicated facilities of extraordinary physical scale.

The costs of these design features varied significantly between urban and rural areas, primarily due to their right-of-way requirements. Not only was land generally more expensive in urban areas, but it often had buildings and other capital improvements that housed businesses or households. The urban Interstates required rights-of-way of sufficient width to build four lanes, on a continuous alignment with acceptable curvature, with adequate space for grade-separated intersections at selected major cross streets.

Generally, these design features increased the constraints engineers faced in siting a particular facility. As a result, it was generally more difficult to avoid existing buildings, monuments, neighborhoods, or other facilities. The higher the design speed, the more difficult it was to design a safe curve as to avoid a particularly important park or monument, because a curve that might be safe at a 30-mph design speed would not be safe at 50 or 60 mph.

The search for acceptable rights-of-way led engineers to consider two categories of land that caused much greater difficulty down the road: parks and slums. Parks were attractive because they were usually already in the public domain and no businesses or houses were located in them. But their preservation later became a rallying cause for highway opponents.

Slums were attractive sites because of prevailing views about the nature of urban "blight," as it was called. Blight was widely viewed like a tumor—cutting it out would prevent it from spreading. The fact that slums housed families of limited means, disproportionately minority, who often had little prospect for finding affordable housing elsewhere, to say nothing of the cost of the move, was not widely considered. Later, of course, the "white man's highway" being built through largely black slums became another rallying cry for Interstate opponents.

The benefits of the design features also varied considerably between urban and rural areas. On the one hand, rural trips are generally longer in distance than urban trips, so speed is at a greater premium for longer trips. The density of traffic and the extent of roadside development, on the other hand, afforded great benefits for limited access in urban areas. Grade-separated intersections protected through traffic from stopping to accommodate entering and exiting traffic. But the cost of grade-separated interchanges limited their financially feasible spacing—which had differential effects for short and long trips.

Such was the optimism of the enthusiasts of the urban Interstates that they assumed that the facilities would be congested only on rare occasions. If the states and localities had built the "complementary" facilities called for by MacDonald and Hilts in their articles, congestion would have been lower. But in any event, the Interstate became a substitute for those facilities, with dramatic consequences from which we are only today beginning to recover.

Limited Mileage

When Congress created the Interstate in 1944, it limited its overall length by statute to 40,000 miles. Such a statutory limitation was consistent with the Bureau's longstanding belief in the need for restricting improvements to a limited system of highways, set forth in the 1921 highway act. It was also consistent with an interregionally oriented system, since it was logically possible to select a network of highways that connected most of the important urban areas and strategically important ports, harbors, and military installations. But a fixed-length system was not consistent with serving the explosively growing metropolitan complexes that began to emerge after World War II.

To select the routes of an interregional system, the states and the Bureau needed to identify major cities and strategically important facilities. Once identified, this set of cities and facilities could be expected to be relatively stable. To be sure, new ports or harbors might emerge due to changing economic conditions or geopolitical considerations, and new cities might grow from less important towns. But the determination of major cities and strategic facilities was unlikely to change much from year to year.

The situation within metropolitan areas was quite different, however, and the effects of the limitation on length were much more problematic. The reason is that, just as MacDonald and Hilts had feared, exclusive development of the urban Interstates without the necessary complementary facilities distorted travel and development. While the Interstate system may have been designed for interregional travel, in metropolitan areas its use was substantially for local trips.

Cost Estimates, Financing, and the Highway Trust Fund

After the statutory creation of the Interstate system in 1944, the Bureau embarked energetically on designating routes and developing cost estimates. In 1945, the Bureau and AASHO adopted the design standards specified in *Interregional Highways*.[31] In 1947, the Bureau and the states had designated most of the routes in the system (excluding urban circumferential highways).

In 1949, the Bureau released its first cost estimate of building the system.[32] The cost estimate was $6 billion; its magnitude was so great that the traditionally unified highway lobby fractured over the issue of how to finance construction. Eventually, it reunified around the creation of the Highway Trust Fund, as specified in the Federal Aid Highway Act of 1956.[33]

Beltways

One of the most significant Bureau initiatives in the designation of Interstate routes was the circumferential freeway or beltway. Many believe that the beltways were originally envisioned as "bypass" routes for traffic bound for destinations beyond the cities they encircled. While they certainly were capable of serving that function, the Bureau viewed them explicitly as measures to contain urban decentralization and to serve city-bound traffic that was not destined for the central business district.

The Bureau had developed the concept of the circumferential or "distributing highway" in *Interregional Highways*. The basis of the concept was that much of the traffic bound for a city was not bound for its central business district, according to traffic surveys. But the radial pattern of most city highways required traffic to pass through the central business district to reach non-central points beyond. Circumferential highways would allow traffic to circulate around the rim of a city and reach its destination from the circumference rather than via its center, thereby, it was hoped, relieving downtown congestion.

---

[31] American Association of State Highway Officials, "Design Standards for the National System of Interstate Highways," adopted August 1, 1945, in *Design Standards: 1-Interstate System; 2-Primary System; 3-Secondary and Feeder Roads*, revised March 15, 1949 (Washington, D.C.), 2-5.
[32] U.S. House Committee on Public Works, *Highway Needs of the National Defense*, 81st Cong., 1st sess. (30 June 1949), H. Doc. 249
[33] "Federal Aid Highway Act of 1956," P.L. 84-627, 70 *Stat.* 374 (29 June 1956). For a detailed discussion of the politics of the debate, see Rose, *Interstate Express Highway Politics, 1941-1989*.

The circumferential concept was also consistent with the Bureau's goal of stopping urban decentralization. The Bureau believed that circumferential highways would facilitate growth in the interstices of the existing radial highway network. By cutting through these interstitial areas, they would provide highway access and facilitate growth close to the central city rather than at the extreme outer fringe.

Circumferential routes were not a new concept in urban road design. Baron Haussmann built his "inner ring" of boulevards in Paris in the latter half of the nineteenth century.[34] Also, railroads had long used circumferential routes for avoiding delays in heavily built-up downtown areas. The circumferential concept, however, was a distinct departure from the "urban extension" concept that had prevailed in the U.S. since the early 1930s. These urban extensions typically penetrated to the heart of the central business district, following old, well-established routes. The circumferential concept departed from this tradition and required construction of many miles of new facilities on new locations, since circumferential routes were not typically extant within the existing network. Furthermore, the scale of the Bureau's circumferential routes was much greater. Haussmann's "inner ring" had a diameter of only two or three miles, while the Bureau's designs called for diameters up to roughly fifteen miles in the larger urban areas.[35] The five-fold increase in diameter would increase the area enclosed within the ring twenty-five-fold.

The designation of circumferential highways proved to be a particularly difficult matter for the Bureau. The process for designating Interstate routes was that the Bureau would request state highway departments to nominate routes for Interstate designation, based on a set of guidelines. The Bureau would then review those nominations, compare the total length of nominated routes from all states, and negotiate with the states to include routes linking particular facilities or cities or to exclude routes it deemed less important. This process was modeled after the designation process used in establishing the primary highway system in the 1920s.

The Bureau's 1946 annual report recounts the difficulty it encountered in convincing the states to adopt the circumferential routes for future expansion.[36] The Bureau's initial split of the 40,000-mile Interstate system allocated 35,000 miles to rural routes and urban extensions and 5,000 miles for circumferential routes. The states' initial submission of routes for designation as Interstate routes totaled 45,070 miles, comprising 43,057 miles of rural routes and urban extensions and only 2,013 miles of circumferential routes. Thus the states' requests exceeded the rural and urban extension quota by 30 percent and fell short of the circumferential quota by 60 percent. Seventeen or eighteen states requested no circumferential mileage at all, despite the Bureau's contention that "... urban conditions in some clearly indicated the need."[37] The Bureau's

---

[34] For a description and evaluation of Haussmann's boulevards, see Howard Saalman, *Haussmann: Paris Transformed*, Planning and Cities (New York: George Braziller, 1971), esp. 14-18.
[35] Saalman, *Haussmann*, plate 29, and *Interregional Highways*, 77.
[36] [U.S. Federal Works Agency], *Work of the Public Roads Administration 1946*, Annual Report (Washington, D.C.: U.S. Government Printing Office, 1947), 9-12. (Reprinted from U.S. Federal Works Agency, *Seventh Annual Report*, 1946.)
[37] The report is unclear as to whether seventeen or eighteen states failed to submit requests for circumferential mileage. Ibid., 11-12.

response to the states' lack of enthusiasm for the circumferential concept was to set aside the designation of the circumferential routes and focus on designating the rural and urban extension portions of the system. This negotiation eventually reduced the length of these portions of the system to 37,224 miles, still above the Bureau's initial allocation of 35,000 miles but substantially below the initial submission.

After resubmitting this reduced system to the states for approval, the Bureau added a few hundred miles to win state acceptance of the system and approved a 37,681-mile system of rural and urban extension routes on August 2, 1947. This final system comprised 34,700 miles of rural routes and 2,882 miles of urban extensions, and left undesignated 2,317 miles for circumferential routes or future expansion.[38] The circumferentials remained undesignated until late 1955, shortly before the enactment of the 1956 highway act.[39]

Thus the beltways around many U.S. cities are explicit products of the Bureau's view of them as the right way to address metropolitan highway demand and counter urban decentralization, or, as we have come to call it, urban sprawl. Once this solution was set into place in the late 1940s and early 1950s, the limitation on Interstate system length and the subsequent financing provisions for Interstate construction locked it in as the dominant American model for addressing urban highway problems.

The consequences for the American urban highway experience were profound. Despite the admonitions of MacDonald, Hilts, and others, most American cities did not build the complementary facilities necessary to balance the investments in the urban Interstates. As highway demand increased, it increasingly utilized the facilities at hand—the urban Interstates—leading to congestion and pressure to expand those facilities much sooner than anticipated. As the scale of the urban extensions and beltways increased to accommodate that traffic, the impacts on the communities they traversed increased as well.

Institutional Arrangements

The institutional arrangements for planning and constructing Interstates were also largely in place by the late 1940s. States would continue to take the lead in rural areas. They would also take the lead in urban areas, and city officials would play a consultative role. Design standards would be aggressively adhered to, even when the impact on local communities was great. A study of state–city relations on highway matters of the time by Wilbur Smith—who later founded a highly successful engineering, planning, and economics consulting practice that

---

[38] [U.S., Federal Works Agency], *Work of the Public Roads Administration 1947*, Annual Report (Washington, D.C.: U.S. Government Printing Office, 1947), 5. (Reprinted from U.S., Federal Works Agency, *Eighth Annual Report*, 1947.)

[39] The locations of the newly designated circumferential routes are illustrated in U.S. Bureau of Public Roads, *General Location of National System of Interstate Highways, Including All Additional Routes at Urban Areas*, Designated in September 1955 [also known as "The Yellow Book"] (Washington, D.C.: U.S. Government Printing Office, 1955).

bears his name—captures the prevailing attitude of state and federal officials about adhering to design standards in cities.

> ... [S]ome city authorities feel that both state and federal standards are so high they tend to retard urban improvements. This view is open to serious question.... [Construction plans and specifications for projects] must meet certain minimum technical standards, as for right-of-way, lane width and other elements of geometric design, and for traffic control devices.... The value of these controls hardly needs discussion. They serve, of course, to insure adequate facilities and adequate performance by the cities. When such standards are considered in their proper light as representing minimum requirements they impose no hardship on the cities. The cities should welcome this check on their performance.... A majority of the states pointed out that sufficient flexibility is allowed in the application of minimum standards to provide for necessary adjustment in special cases, particularly with respect to right-of-way requirements.... At this stage of development a certain amount of education is still needed.... [S]ome cities claim the states' standards are too high, especially for lane width and right-of-way. One state official put the case very well in saying that the state standards may sometimes be difficult to attain but they are not too high. Certainly too many glaring examples of past shortsightedness confront us to risk similar errors by failure now to adopt adequate design standards.... Design and other technical standards should be based on best practice, and in light of even today's experience it would be far better to err on the side of high standards.[40]

This paternalistic attitude about design standards as painful but necessary medicine for the cities led to considerable conflict. Cities often wanted facilities that would impose minimal impacts. The states, with support from the Bureau, often insisted on condemning sufficient right-of-way to build facilities to standard. But their subordinate position in decision making often gave them little ability to control design decisions, which led to major conflict down the road.

By the early 1950s, then, the stage was set for a showdown over urban Interstates, even though it did not begin in earnest for another decade. Most rural areas were subject to much less conflict. Right-of-way was at much less of a premium, and dislocations to houses and businesses were much less acute. Indeed, the rural Interstates were viewed as a source of access and prosperity in many quarters. The exception was highways through scenic and environmentally sensitive areas. Here, too, the design standards became a source of conflict and delayed construction, sometimes for decades.

***

The first half-century of the American highway program transformed a fragmented, disjointed collection of county roads to an integrated, interconnected national system of highways. Yet the first generation of highways it produced were not capable of accommodating the explosion of road traffic that followed World War II. The Interstate system, the planning for which began

---

[40] Norman Hebden and Wilbur S. Smith, *State–City Relationships in Highway Affairs* (New Haven: Yale University Press, 1950), 45-46, 49-50, 61, 121, 189-90, 211.

before the war, finally received the financing necessary for construction in 1956. It promised a solution to a country clamoring for highway improvements. Alas, its implementation, while it yielded many of the benefits it promised, also prompted a reexamination of national priorities for the preservation of communities, air quality, and natural resources, which are the subject of the next chapter.

4

THE AMERICAN HIGHWAY PROGRAM SINCE 1956

On June 26, 1996, in a huge tent on the Ellipse in front of the White House, Vice President Al Gore commemorated the fortieth anniversary of the Interstate highway system by honoring four Americans—"The Visionaries"—who played a central role in its creation. First was President Dwight David Eisenhower, who signed the 1956 bill that created the Highway Trust Fund and effectively launched the construction of the Interstate system. Many years earlier, in 1919, Eisenhower had led a military convoy from Washington to the Pacific that required more than two months and exposed the deficiencies of the national highway system of the day. His granddaughter Susan Eisenhower accepted his award.

Second was Congressman Hale Boggs, who led the crusade for the Trust Fund in the House of Representatives. In 1972 he disappeared in a small plane over Alaska and was never found. His widow, Lindy Boggs, accepted his award. She had succeeded him in the House for eighteen years and later became the U.S. Ambassador to the Vatican.

Third came Frank Turner, who started working in the Bureau of Public Roads in 1929 and rose to become the Federal Highway Administrator before retiring in 1972. And finally there was Senator Albert Gore, Sr., the vice president's father, who had played an important role in the Senate's passage of the bill. Award recipients received crystal replicas of the "Zero Milestone" monument on the Ellipse, dedicated by President Warren G. Harding in 1923 at the originating point of Eisenhower's convoy.[1]

It was a classic Washington event: bipartisan, highly produced, and complemented by a national public relations campaign to underscore the urgency of reauthorizing the federal highway program the following year. The American Highway Users Alliance released a report for

---

[1] American Highway Users Alliance, *Celebrating America's Highways: A Commemorative Scrapbook* (Washington, D.C., 1996).

the anniversary entitled "The Best Investment a Nation Ever Made."[2] That same month the U.S. secretary of transportation retraced part of Eisenhower's journey, with appearances in key cities along the route. So great was the hoopla that a reform group, the Surface Transportation Policy Project, used the occasion to launch a campaign of letters-to-the-editor to showcase the damage done by the Interstate program.

The forty years between 1956 and 1996 saw a remarkable transformation in the climate for highways in the U.S. Designing and constructing the Interstates, and stopping and modifying them, was the dominant federal surface transportation policy issue until 1991—for three and one-half decades. It was a fascinating period in American surface transportation policy that has only begun to be understood.[3]

BUILDING THE INTERSTATE: 1956-1991

The design and construction of the Interstate was one of the major engineering enterprises of the twentieth century. The physical magnitude of the project was enormous. It taxed the engineering capacity of the U.S. It was the largest civil works project in memory. The impacts of a system so large are incalculable in many respects. It will be hard to sort out its long-term impacts until we can look back over decades.

A thorough analysis of the period would require a book or several books. The overview presented here highlights events and developments relevant to this book's focus on policy and decision making for future American transportation policy.

The 1956 Highway Acts

The fortieth anniversary celebration of 1996 was in fact the anniversary of a funding mechanism, the Highway Trust Fund. The Interstate system itself had been the subject of planning and analysis since the late 1930s and was legally established in 1944. After 1956, construction of the Interstate became a crash program. The statute called for constructing the entire system, by now 41,000 miles in length, in only twelve years.[4]

The 1956 legislation brought fundamental changes in federal highway financing. Prior to 1956, federal highway spending came from general revenues; after 1956, it came from a dedicated

---

[2] Wendell Cox and Jean Love, *The Best Investment a Nation Ever Made: A Tribute to the Dwight D. Eisenhower System of Interstate and Defense Highways*, <http://home.i1.net/~policy//tp-is40.htm> (Washington, D.C.: American Highway Users Alliance, June 1996).
[3] See, for example, American Association of State Highway and Transportation Officials, *The States and the Interstates: Research on the Planning, Design and Construction of the Interstate and Defense Highway System* (Washington, D.C.: The American Association of State Highway and Transportation Officials, 1991); Larry Hott and Tom Lewis, *Divided Highways*, Film (Florentine Films, 1997); Tom Lewis, *Divided Highways: Building the Interstate Highways, Transforming American Life* (New York: Viking, 1997); Mark H. Rose, *Interstate Express Highway Politics, 1941-1989*, rev. ed. (Knoxville: University of Tennessee Press, 1990).
[4] The Federal Aid Highway Act of 1944 (58 *Stat.* 838, 20 December 1944) created the Interstate; the Highway Revenue Act of 1956 (70 *Stat.* 387, 29 June 1956) established the Highway Trust Fund, and the Federal Aid Highway Act of 1956 (P.L. 84-627, 70 *Stat.* 374, 29 June 1956) authorized the spending program.

source, the Highway Trust Fund. The trust fund received earmarked income from federal excise taxes, principally a gasoline tax of three cents per gallon. The previous two-cent gas tax had gone into the general fund. Hence, after 1956 the federal highway program had a revenue source that was protected from the year-to-year budgetary struggles that faced general fund programs, all of which had to compete with each other for funds.

On the spending side, the 1956 acts increased the federal share of Interstate construction to 90 percent from the previous level of 50 percent. This provision dramatically changed the economics of Interstate construction for the states. Whereas they had previously had to match federal-aid dollars one-for-one, states now needed to contribute only ten cents for each dollar of Interstate construction expenditure. In western states with large areas of federally owned lands, the federal match could go as high as 95 percent.

This provision virtually eliminated the attractiveness of toll roads, which had been growing in popularity among the eastern and midwestern states since the end of World War II. Under the traditional fifty–fifty division of costs between the state and federal governments, toll roads had the attractive feature of providing superior service at no budgetary cost to the state government. The cost of expressways was so great that even paying half the cost of a facility could absorb a significant portion of the state highway budget. With a state paying only 10 percent of the total cost, however, toll financing and the cost of interest became much less attractive. Moreover, the economic multiplier effects of Interstate construction could arguably return the state's match through increased state tax revenues. Not surprisingly, few toll road projects were initiated in this period, and superhighway construction became almost exclusively the province of the Interstate program.

A second important feature of the 1956 act was its provisions for allocating funds to the states from the Highway Trust Fund. Traditionally, federal-aid funds were apportioned to the states according to a formula weighted equally among three factors: roadway mileage, land area, and population. The 1956 act apportioned funds to the states instead by a formula based on the cost to complete the Interstate in each state. Thus, for example, if the cost to complete a state's Interstate routes was 2 percent of the cost to complete the entire Interstate system, then that state would receive 2 percent of the total Highway Trust Fund apportionment for Interstate construction in a given year. The state could then draw on that apportionment to reimburse 90 percent of its expenditures for constructing Interstate routes.

The "Golden Age"

The states responded quickly to the availability of the new funds and to the national initiative to build the system. Within weeks of the signing of the act into law on June 29, 1956, states had begun to let contracts for construction, and by November the first section of highway was open.[5]

---

[5] Richard F. Weingroff, "Three States Claim First Interstate Highway," special issue of *Public Roads* (June 1996): 17–18.

52 FLEXIBLE URBAN TRANSPORTATION

But state highway departments also felt the crunch that such a large program would bring. The construction of the system would require vast resources in terms of materials, engineering effort, administration, and labor. State highway department employees of the period stress the technical and administrative challenges of gearing up for such a large program. But they also stress the widespread enthusiasm for the task in many states. And even states with relatively primitive highway systems in the 1950s, such as Mississippi, geared up quickly to start designing and constructing the Interstates.[6]

Construction proceeded rapidly on much of the system, although not as rapidly as originally anticipated. By 1969, the original target year for completion, only 25,650 miles were open for operation, 63 percent of the originally authorized 41,000 (see Figure 4-1)

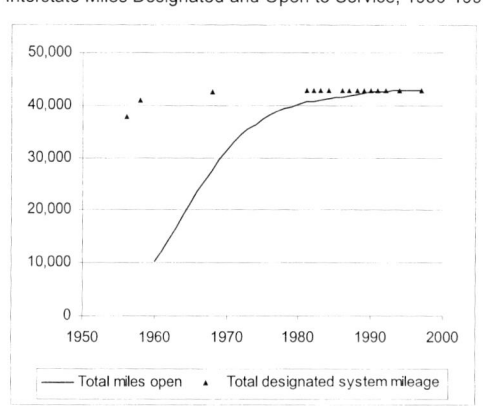

Figure 4-1
Interstate Miles Designated and Open to Service, 1956-1997

Source: U.S. Federal Highway Administration, Office of Program Administration, personal communication (March 6, 2001).

The slower than anticipated progress on construction derived in part from cost increases and the "pay as you go" provisions of the Highway Trust Fund. The estimated cost to complete the system increased from $41 billion in 1961 to $89 billion by 1975 (see Figure 4-2). Cost increases derived from several sources: unit costs increased,

Congress increased the minimum design standards for the system, and Congress increased the length of the system to 42,500 miles in 1968 (see Figure 4-1). Controversy over the impacts of the Interstate in cities and in environmentally sensitive areas began to delay projects as well.

By the mid-1960s, then, the crash program of 1956 had made much progress in some areas. In some city and rural locations, however, the Interstate had begun to bog down. The opposi-

---

[6] American Association of State Highway and Transportation Officials, *The States and the Interstates*, 30-57.

tion came as something of a shock to the highway community, which had traditionally enjoyed widespread support. Indeed, most controversy had been about who would get the limited highway funds that were available.

The opposition had a profound effect on the highway program. Its decision processes had traditionally been largely based on consensus. Now it would face a welter of legal and procedural obstacles.

The Freeway Revolt

The "Golden Age" of highway building was relatively short-lived. The ambitious transportation plans developed in the 1950s and early 1960s ran into major opposition in many American cities and in environmentally sensitive rural areas. As the Interstate began to penetrate cities, opposition to its design and construction began to emerge. The "Freeway Revolt" caught fire, first in California, then in Boston, and then in many American cities.

Figure 4-2
Estimated Cost to Complete the Interstate, 1958-1975

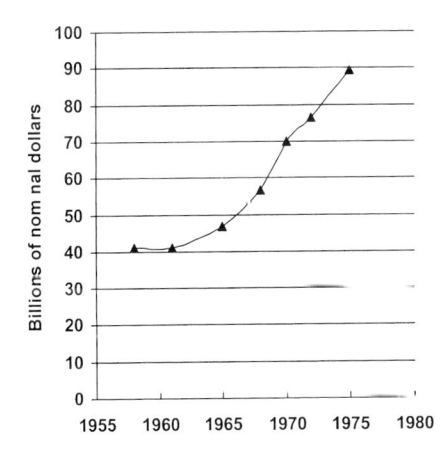

Source: U.S. Federal Highway Administration, *America's Highways, 1776-1976: A History of the Federal-Aid Program* (Washington, D.C., 1977), p. 484.

Five important sources of opposition emerged during the 1960s. First was the impact of urban Interstate design on neighborhoods and open space. Across the country, highway planners had located major highways that cut right through neighborhoods and parks. While this may have seemed reasonable to the planners at the time, it soon became apparent that the public would simply not accept such impacts.

To its opponents, the Interstate program was destroying American cities in the interest of the automobile. In their view, highways were laid out to provide a high-speed alignment to carry suburban workers into the central city to work, with almost no regard for the preservation of parks, scenic vistas, or neighborhoods. Their opposition was exacerbated by the fact that, frequently because of the supposedly interstate nature of the traffic to be served, on-ramps and exits from the system were separated by fairly long distances in a manner that reduced their usefulness for local trips. As a result, the highways were less useful to those on whom its negative impacts fell most heavily.[7]

Second, public sentiment was giving increasing importance to the environment. The late 1960s and early 1970s saw the enactment of sweeping new federal statutes controlling the pollution of air, water, and land. Projects receiving federal money—like most highways—were subject to heightened scrutiny. The National Environmental Policy Act of 1969 (NEPA) required environmental impact assessments for every project. Projects could proceed only if they won out as the most reasonable choice from a broad range of carefully considered alternatives. Moreover, the conclusions of the assessments, and the procedures followed in conducting the assessments, were subject to challenge in the courts. Any group that disagreed with the outcome of an assessment could take the planners and the construction agencies to court, halting a project in its tracks. Other major environmental legislation included the Clean Air Act Amendments of 1970 and the Noise Control Act of 1972.[8]

A third source of opposition grew out of the Civil Rights movement. Planners of the 1950s had ideas about the sources and remedies for poverty that differ sharply from the dominant views today. At that time, slums were seen as "blight," like cancer, that would spread to other areas if not removed. The need to locate new highways, coupled with this view of disadvantaged neighborhoods, led to the location of many the urban highways of the time through poor neighborhoods that were often predominantly black. Indeed, some of the first highways built were in poor neighborhoods. Since many of the residents rented their dwellings, the highway building agency tended to compensate the landlord for the "fair market" price of a property, but let the residents shift for themselves to find new apartments. Often the construction agencies gave renters only a token payment for moving expenses.

The impact on some neighborhoods was devastating. Displaced renters often faced great difficulty in finding new apartments. Racial bias in housing made the problem worse. Some leaders in the growing Civil Rights movement began to see the highway program as an instrument of a racist majority, building the "white man's highway through the black man's neighborhood."

A fourth source of opposition was the oil embargo of 1973 and the ensuing "oil crisis," which led many to conclude that continued dependence on automobiles would endanger national security. Rationing, long lines at gas stations, and government price controls made matters worse.

---

[7] For a discussion of this issue in the San Francisco Bay Area, see William H. Lathrop, Jr., "The San Francisco Freeway Revolt," *Transportation Journal of the ASCE* 97 (February 1971): 141-42.
[8] "National Environmental Policy Act of 1969," *Stat.* 82 (1 January 1970): 852; "Clean Air Act Amendments of 1970," P.L. 91-604, *Stat.* 84 (31 December 1970): 1676-713; "Noise Control Act of 1972," *Stat.* 86 (7 October 1972): 1234.

A final cause of the Freeway Revolt was a widespread loss of trust in large government institutions. The Vietnam War (1954-1975) and Watergate (1972-1974) made many people skeptical about the conclusions of government "experts" like transportation planners and engineers. The result was a crisis in administrative authority in American government. The clean, technically based scientific management of the Progressive Era was downgraded in favor of community-based participatory planning and decision making.

The legislative development of the highway program tracked these larger social movements closely. The Federal Aid Highway Act of 1973 allowed for the first time the exchange of Interstate highway funds for mass transit funds, allowing cities to eliminate urban sections of the Interstate system and apply the money thereby liberated to the development of urban transit systems. Hard-fought battles in both rural and urban areas raised the status of environmental criteria in the decision-making process. Highway location and design based on the application of technical criteria gave way to community-based and political criteria.[9]

Process, Design, and Values

As the Freeway Revolt unfolded during the 1970s, it precipitated an identity crisis in the American highway community. To the engineers who were building these highways, the Interstates offered the fruits of technological progress and Progressive enlightenment to American cities. Instead, they found their offering repudiated and themselves repudiated as exponents of an evil, destructive technology. Professionals of the era speak in painful terms of how the conflict over the Interstates precipitated crises in their own careers.[10]

It is important to recognize that the Freeway Revolt grew out of a combination of concerns about process, design, and values. The procedural concerns related to who made decisions about the necessity, location, and design of Interstates. The design concerns related to the scale of facilities—primarily the effect of capacity, grade separation, and design speed—and the spacing and location of access points. Concerns about values related to the role of autos and transit in American cities, the primacy of environmental concerns, and the preservation of traditional cities.

Resolving these concerns and conflicts dominated the surface transportation agenda for the remainder of the Interstate era. The result was a fundamental restructuring of the process for making location and design decisions about highways. The nature of the transformation is illustrated well in thumbnail sketches of three controversial projects: Interstate 66 in the northern Virginia portion of the Washington, D.C., metropolitan region; Interstate 70 through Glenwood Canyon, Colorado; and Interstate 93 through Franconia Notch, New Hampshire.

---

[9] "Federal Aid Highway Act of 1973," *Stat.* 87 (13 August 1973): 250.
[10] American Association of State Highway and Transportation Officials, *The States and the Interstates*, 58-77.

Interstate 66, Fairfax and Arlington Counties, Virginia

Interstate 66 (I-66) is one of the major highways in the Washington, D.C., metropolitan region. Its development took place against a backdrop of controversy in the region about the character of its transportation system. The end result was a highway with significant departures from Interstate standards, both in design and operating characteristics.

The Washington metropolitan region has a long and distinguished history of long-range planning, from the L'Enfant Plan of 1791 and the "City Beautiful" McMillan Plan of 1901. The region grew remarkably during the Depression and World War II, as the federal government drastically expanded its size and responsibility to address those national emergencies. After World War II, the region began to develop plans for its highway and transit systems to accommodate and shape the growing demand for cars and for suburban residences.

In 1961, the National Capital Planning Commission (NCPC) and National Capital Regional Planning Council (NCRPC) published *A Policies Plan for the Year 2000: The Nation's Capital* (referred to hereinafter as the Year 2000 Plan). It was the product of an exhaustive comprehensive planning process, driven largely by the federal government, with an ambitious and challenging forty-year perspective. Its underlying concept was to channel growth into defined growth corridors that would radiate, starlike, outward from the central core, with wedges of open space in between. High-volume, high-speed transportation facilities—rapid rail and freeways—were to be the spines of the growth corridors. Growth centers—communities of varying size and function—were to be located along defined corridors.[11]

The centerpiece of this concept was the proper location and timely development of the rapid rail lines and freeways. Rapid rail lines and freeways were seen as key leading land uses, constructed in the corridors and not in the wedges. Five years later, in 1966, the NCRPC published *The Regional Development Guide 1966-2000,* a more detailed policy document that updated, refined, and modified the Year 2000 Plan while retaining its basic outline. In the intervening years, the Maryland–National Capital Planning Commission adopted *On Wedges and Corridors: A General Plan for the Maryland–Washington Regional District* (1964), and the Northern Virginia Regional Planning and Economic Development Commission released its *Northern Virginia Regional Plan: Year 2000* (1965). Both followed the principles of the Year 2000 Plan.[12]

The 1961 Year 2000 Plan called for six growth corridors interconnected by three circumferential freeways: an Inner Loop circling the core, the present-day beltway (comprised of Interstate 495 and a segment of Interstate 95), and an outer circumferential. In 1966, NCRPC's *Regional*

---

[11] National Capital Planning Commission and National Capital Regional Planning Council, *A Policies Plan for the Year 2000: The Nation's Capital* (Washington, D.C.: 1961).
[12] National Capital Regional Planning Council, *The Regional Development Guide 1966-2000,* prepared by the Comprehensive Planning Division (Washington D.C., June 30, 1966); Maryland–National Capital Park and Planning Commission, *On Wedges and Corridors: A General Plan for the Maryland–Washington Regional District* (June 22, 1964); Northern Virginia Regional Planning and Economic Development Commission, *Northern Virginia Regional Plan: Year 2000* (Arlington, Va.: 1965).

Development Guide expanded this concept to seven corridors and added a fourth beltway. Outlying circumferentials were to interchange only with the corridor freeways. This policy had two purposes: to strengthen the corridor and to prevent development pressures within the open space wedges.

The 1961 Year 2000 Plan was developed on the premise of a strong, unified governmental role, dominated by federally chartered regional institutions with sufficient authority to build the facilities they planned and to impose the controls and limitations on development to make the plans work. At the time, an assumption of federal dominance may have been reasonable. Senator Alan Bible was chairing a Joint Committee on Washington Metropolitan Problems, which was actively dealing with a full range of issues: growth and expansion; water supply, pollution, and sewage disposal; transportation; planning; economic development; and governance. Out of this substantive federal interest emerged calls for a National Capital Metropolitan Conference, a Federal Regional Development Agency, and the National Capital Transportation Agency (NCTA).

Yet strong regional institutions have had difficulty developing. Indeed, one of the institutions with regional scope has lost much of its power. The establishment of home rule in the District of Columbia in 1967 significantly eroded the influence of the National Capitol Planning Commission, which previously had considerable influence to approve or veto actions that affected federal buildings and other installations. Today, the Metropolitan Washington Council of Governments is the primary regional entity, and its Transportation Planning Board acts as the official regional transportation planning entity (officially, its metropolitan planning organization, or MPO).

The region is primarily governed by three sovereign entities, the District in the center and Maryland and Virginia, each governed from remote state capitals. The District has focused much of its energy over the last three decades on establishing its own democratic institutions after a century of leaders appointed by the president. The Virginia suburbs have struggled under the limitations of the Dillon Rule, which restricts their authority to act on any issue without express permission from the state legislature, and state authority over most roads. Maryland suburbs are much more autonomous from their state capital than Virginia. They own their own roads and are therefore much less tied to the state transportation department.

As the national Freeway Revolt played out in the Washington, D.C., metropolitan region, many of the highways planned in the 1960s literally disappeared from the plans. Public concern about air quality, civil rights, petroleum dependency, and neighborhood preservation rendered some of the proposed highways untenable. Popular support turned to public transit, although the private automobile, then as now, remains overwhelmingly the personal mode of choice. The prevailing attitude against freeways, coupled with the opportunity to divert federal funds from freeway construction to the building of a heavy rail metro system, caused political support for freeway construction to all but evaporate.

As a result, many of the planned highways were canceled. Approximately 244 miles of freeway for the major routes and almost 1,500 lane-miles of freeway were planned but not constructed (see Table 4-1). Compared to 311 miles of freeway in the region today, and 1,914 lane-miles, the Washington Freeway Revolt trimmed almost half of the planned freeway capacity from the region.[13] Other highways were delayed considerably, and in some cases scaled down in design from limited access freeways to wide arterial streets with at-grade intersections and traffic signals. By the mid-1980s, few of the unbuilt highways from the 1960s' plans even remained in the long-range plan.

Table 4-1
Projects Deleted from the Washington Freeway Plan, 1961-1996

Project	Lanes	Center Line Length (mi.)	Lane Miles
District of Columbia			
North Leg Freeway	6	3.5	21.0
South Leg Freeway	6	1.9	11.4
East Leg Freeway	6	4.3	25.8
Industrial Freeway	6	3.1	18.6
Potomac Freeway	6	1.2	7.2
Palisades Parkway	6	1.5	9.0
North Central Freeway	6	6.0	36.0
Subtotal		21.5	129.0
Maryland/Virginia			
Second Beltway	6	99.0	594.0
Outer Beltway	6	70.0	420.0
Bluemont Expressway	6	6.0	36.0
Four Mile Run Expressway	6	6.0	36.0
Monticello Freeway	6	22.0	132.0
Pimmit Expressway	6	5.0	30.0
Potomac Freeway	6	15.0	90.0
Subtotal		223.0	1,338.0
Total		244.5	1,467.0

Source: Greater Washington Board of Trade, "History and Current Conditions," report 1, 1997 Regional Transportation Study (February 1997), p. 3.
Note: Initial plans for some suburban freeways assumed four to six lanes. This calculation assumes right-of-way for at least six lanes.

The controversy over Interstate 66 occurred against this backdrop of regional concern about freeway expansion. Interstate 66 serves the traffic corridor between Washington, D.C., and

---

[13] U.S. Federal Highway Administration, *Highway Statistics 1994* (Washington, D.C., 1995), tables HM-71, HM-72.

northwestern Virginia, connecting at its western end with Interstate 81. It was constructed in two phases. The section from the Washington Beltway to the western terminus at I-81 was constructed early on in the Interstate program. The more controversial section ran from the beltway around Washington, D.C., into downtown. It now serves a high volume of daily commuting traffic from suburban Virginia west of Washington.

The Virginia highway department originally proposed an eight-lane, at-grade facility from inside the beltway to just outside the District, where it would have split to meet the various crosstown expressways and inner city ring roads envisioned in the Year 2000 Plan. The proposed route passed through heavily developed residential and commercial sections of Arlington County, and the community expressed great concern about the impact of such a large facility built at-grade. Years of litigation and injunctions followed. Eventually, a cooperative design approach was adopted that included significant community involvement. The U.S. secretary of transportation approved the final environmental impact statement on January 5, 1977, and five years later, in 1982, the entire 10-mile facility was open to traffic.

The highway that was constructed differed in some significant aspects from typical urban Interstates of the time. It was limited to four lanes, rather than the eight originally proposed. Highway geometry is full Interstate standard for most of the length. The design speed is 60 mph, which is the urban standard, for all except a short section in the town of Rosslyn, where it is 55 mph. Lane widths are 12 feet, right-hand shoulders are 15 feet, and inside shoulders are 8 or 12 feet. The Washington Metro line is in the median for several miles (from Glebe Road in Arlington to Nutley Street near Vienna, about two miles beyond the beltway). The highway is depressed for 2.5 of its 9.6 miles inside the beltway, and is extensively landscaped. It includes a park that is built over the freeway in one section. Sound barriers are provided along most of the length. And a path for bicycling and jogging along the entire length has been incorporated into the right-of-way.

The below-grade sections of the highway in Arlington, based on personal observations by the author, have dramatically reduced its impact on the surrounding community. Many local streets continue without interruption at grade across the depressed highway.

The agreement that allowed I-66 to go forward also included significant operating restrictions. High occupancy vehicle (HOV) restrictions are in effect for all inbound lanes during the morning rush and for all outbound lanes during the evening rush, excluding single-occupant vehicles for downtown commuting along this route during rush. Over the years since it opened, the definition of high-occupancy vehicles has relaxed until it is today a minimum of two persons. Trucks are prohibited at all times, making this highway functionally more like a parkway than an Interstate highway.

The design and operating restrictions on I-66 have also had an impact within the state highway department. Then-commissioner Douglas Fugate is said to have commented that 66 convinced him that if northern Virginia did not want additional highways, the state highway department would simply shift its resources to areas of the state that did. He was concerned that many years and a great deal of money had been spent to build a facility that had too little capacity and poor operating characteristics.

Interstate 70, Glenwood Canyon, Colorado

Interstate 70 (I-70) west of Denver was not part of the original 40,000-mile system authorized in 1944. The project had been proposed early in the century and was added to the Interstate system as part of the 1,000 additional miles authorized in the 1956 act after intense pressure from Colorado. On technical grounds it was thought to be the only feasible route over the Rocky Mountains west from Denver. Like I-66, the Glenwood Canyon project resulted in significant departures from Interstate standards, in this case due to a desire to mitigate environmental impacts on popular scenic and recreation destination.[14]

The Rockies had divided the western part of the state from Denver and the rest of the state. In fact, in the early years of aviation, one could only fly from Denver to Grand Junction by going north to Cheyenne, Wyoming, or south to Colorado Springs. Road passage over the mountains in winter was virtually impossible. An existing two-lane road was subject to severe congestion during the peak recreational season. Glenwood Canyon was a popular recreational destination for hikers, and the western slope of the Rockies was at that time heavily favored for the development of downhill skiing.[15]

After designation in 1957, work on I-70 proceeded with a relatively low profile. Public hearings on the improvement of four miles of Interstate east of Glenwood Springs, including twin tunnels near No Name Creek, were held in July 1963. Another public hearing in March 1964 covered I-70 through Glenwood Canyon. When completed, the section of I-70 near No Name Creek had talus slopes and cut scars that were permanent and considered by some to be environmentally damaging, which raised public concern over the environmental impact of building an Interstate through Glenwood Canyon. In late 1968, the Colorado legislature passed a joint resolution calling for the creation of a Citizen Advisory Committee (CAC) to advise the Colorado Department of Highways (CDOH) on the Glenwood project. In early 1969, the governor appointed the committee, comprised of representatives from a variety of professional disciplines, including architects specializing in planning, landscape, and structures.

The CAC met four times, toured the route on foot, and recommended in June 1970 that CDOH study other route alternatives, conduct a corridor public hearing, and cease right-of-way purchases until the route was finalized. CDOH objected to the tone of these requests as failing to recognize CDOH's authority in these matters. CDOH informed the committee that other routes were being investigated, that a corridor public hearing would possibly be held in October, that right-of-way purchases were a CDOH prerogative, and that "it would appear that there will be no

---

[14] This case study is adapted from an earlier version in American Association of State Highway and Transportation Officials, *The States and the Interstates*, 97-99.
[15] Charles Shumate, interview, 1988; Marion C. Wiley, *The High Road* (Colorado Department of Highways, Division of Highways, 1976), 61-66.

need for further meetings of the committee until such time as the routing is determined."[16] The committee apparently did not meet again.

In August 1970, CDOH retained a consultant to conduct a location study of the major alternatives to the Glenwood Canyon corridor. Their recommendation was a four-lane highway through Glenwood Canyon, which the department submitted to the Federal Highway Administration (FHWA) in March 1972. FHWA recommended approval in September 1972, but after considering it for six months, the assistant secretary of transportation for environment in March 1973 denied the approval and asked for further analysis of route alternatives.

The highway department had proposed a least cost design utilizing extensive cut-and-fill that would have filled much of the base of the canyon with highway embankment. Such a facility would have significantly disrupted the environment of the canyon. A stream that ran through the canyon would have been rerouted through culverts. Wildlife movements across the canyon would have been seriously limited, and the visual impact would have been significant.

At about the same time, a group of concerned private citizens made a movie advocating that CDOH look at alternatives to its largely cut-and-fill design. In particular, it advocated the use of elevated design concepts, in use in Europe, that virtually eliminated cut and fill by building the road on top of concrete columns. This elevated design minimized disruptions to the ground surface, as well as drainage and animal migration patterns.[17] Charles Shumate, then head of CDOH, opposed a shift to a new design concept on the grounds of cost.

Two FHWA approvals were needed to proceed with the project: corridor location and design. Concern over the design that would be used if the Glenwood Canyon corridor were chosen led environmentalists to try to link the two, whereas CDOH wanted to proceed with approval of corridor location and design in serial fashion. CDOH's assessment was based on the fact that the Glenwood Canyon corridor was less expensive and, in their view, had less of an environmental impact because alternative routes would cut through relatively undeveloped territory, whereas Glenwood Canyon already had a highway and a railroad.

After being denied approval by U.S. Department of Transportation (DOT), CDOH and FHWA began again to build the case for a Glenwood Canyon corridor. Upon the advice of FHWA, the governor, in September 1973, appointed a new committee to review the alternative corridors and recommend whether further study of alternatives other than Glenwood Canyon was required. The committee studied the issue during October, and in November recommended that no further studies of the alternatives were needed to inform a corridor selection. In December, CDOH forwarded the committee's recommendation to FHWA, and again requested approval of the Glenwood Canyon corridor.

---

[16] Wiley, *The High Road*, 62.
[17] *Glenwood Canyon, Colorado*, 35-mm film, available from Glenwood Business Center, Glenwood Colorado (1972).

In the meantime, CDOH had received approval for and retained three consultants to develop design alternatives for the Glenwood Canyon corridor. These were presented in public meetings in July 1974 and forwarded with a third request for corridor approval in August. In December, CDOH submitted for consideration a fourth design concept that combined the best features of the three developed by the consultants.

In January 1975, newly elected governor Richard D. Lamm took office and expressed a desire to move Glenwood Canyon off dead center. He expressed a commitment to the highway, but not necessarily with full Interstate design standards, and expressed his desire for an exemption from the standards for the project.

In August 1975, only two weeks after Charles Shumate retired as executive director of the state department of highways, Secretary of Transportation William T. Coleman, Jr., wrote Lamm that the corridor could be approved if an acceptable design was provided, adding that he would approve nothing less than four lanes. Finally, in December 1975, Secretary Coleman gave preliminary approval and sent the environmental impact study (EIS) on to its final hurdle, the Council on Environment Quality, which granted approval in February 1976.

In the meantime, Colorado senators Floyd K. Haskell and Gary W. Hart introduced an amendment to the highway authorization bill that would exempt Glenwood Canyon from the statutory requirement for four lanes and allow the secretary to waive other design standards as well, in the interests of preserving the scenic beauty of the canyon and improving the safety of the existing facility. The amendment was unopposed and uncontroversial, and had the support of Governor Lamm and Secretary of Transportation Coleman.[18]

Thus, in February 1976, CDOH finally had corridor approval, as well as an exemption from the normal governing design standards for Interstate highways. But ten years had passed since FHWA had authorized the beginning of preliminary engineering estimates in June 1966.

The next task was design. CDOH developed an open process, explicitly intending to consider environmental and aesthetic matters with maximum participation from the public and affected agencies in order to arrive at a consensus solution. The CAC was restructured so that it included representatives from the general public suggested by the counties involved, as well as representatives suggested by environmental and pro-development groups and by the American Institute of Architects. A technical review group (TRG) consisted of representatives from federal and state agencies that had an interest in the project.

In December of 1976, the CAC reported eighteen recommendations covering environmental issues, recreational potential, access, highway geometry, and opinions on design concepts. They recommended study of a two-lane alternative, although they felt that a four-lane highway would be possible.

---

[18] *Congressional Record*, 121 (1975), p. 40014.

The design team, along with CDOH, FHWA, CAC, and TRG, reached consensus on a design and scheduled a public hearing in March 1978. After reviewing comments, they issued a draft design report in October 1978. Although not required, the design report was released for public comment, yielding 818 positive and 131 negative responses. The design report was submitted to FHWA in March 1979, which approved it in September 1979 and authorized the preparation of the final design. Construction began in January 1980, and final construction was completed in 1992.

The completed highway has four lanes, with a 50-mph design speed, 12-foot lanes, 3.5-foot inside shoulders, and 6- or 8-foot outside shoulders. Thus, the significant departures from Interstate standards are the design speed and the shoulder widths. It is located largely on embankment with retaining walls and on structures, with two tunnels (4,000 and 600 feet). Recreational facilities consist of three full rest stops, vehicle access to two other recreational areas, and a continuous bike and pedestrian path. Extensive revegetation of areas damaged by earlier construction was also provided.

Since its completion, the Glenwood Canyon project has been hailed as a design triumph. It has received more than thirty design awards, including the 1993 Outstanding Civil Engineering Achievement Award from the American Society of Civil Engineers and the 2001 Presidential Award for Design Excellence.

In the view of environmentalists, the Glenwood Canyon is a textbook case of sensitizing a highway department. The advisory committees were imposed on the department, first by the state legislature, then by FHWA. Some of the department's own employees becoming strong defenders of environmental concerns. The design was a joint effort, and it incorporated many departures from Interstate design standards. With extensive involvement of community and environmental interests, a new process emerged that is a model for projects in environmentally sensitive areas.

In the view of some critics, the Glenwood Canyon experience is a textbook case of compromising cost effectiveness for marginal environmental utility. What was originally proposed would have cost $18 million. What was finally built cost over $400 million. Highway officials did not suddenly "discover" the environment, according to this view, they were required to spend enormous sums of taxpayer money to provide recreational and scenic amenities to a privileged few. The widespread favorable reception of the Glenwood Canyon project has largely quieted this view, however.

Franconia Notch, New Hampshire

The third case involves the construction of Interstate 93 through the White Mountains in New Hampshire, near one of the state's most popular tourist and scenic destination, Franconia Notch. The existing highway through the park became heavily congested during the recreation season. This led to hazardous road conditions, especially when heavy trucks mingled with recreational traffic, leading frustrated drivers to attempt dangerous passing maneuvers. Frequent roadside parking and pedestrians on the highway led to concern about pedestrian safety and accidents.

The route through Franconia Notch appeared to be the only technically feasible alternative. Controversy over the designation of an Interstate through the Notch dated from the earliest Interstate designation discussions in 1948. The legislature accepted the designation of I-93 through the Notch only in 1959, but stipulated a "parkway like" highway.

The highway department held its required public hearing on the project in late March 1966 (still winter in that region), a venue questioned by some who felt that a project affecting an attraction popular statewide should have been given at a more accessible place and time. The state highway department proposed a conventional Interstate design with a cross-section of four 12-foot lanes, 10-foot shoulders, and a 30-foot clear cut from the roadside to remove roadside obstructions for safety purposes.[19]

An ad hoc group of citizens concerned about the impact on the park appealed to the governor, who created an advisory committee on the project, chaired by former state legislator, congressman, governor, and presidential chief of staff Sherman Adams. Park advocates argued for a more flexible design that was more readily fitted to the particular environment of Franconia Notch. The conventional design, according to their reasoning, would be extremely damaging to the character of Franconia Notch. Another concern was that the blasting necessary to build a conventional cross-section would harm a geological feature known popularly as the Old Man in the Mountain, a cliff face that from certain perspectives resembled the face of an old man.

The conflict wore on and environmental proponents used the provisions of the National Environmental Policy Act (NEPA) to delay and eventually enjoin construction of the controversial 10-mile segment in 1974. Special legislation was introduced into Congress that year to allow deviation from Interstate standards requirements for four 12-foot lanes through this area.

Finally in 1977 an agreement was reached, and with amendments to the agreement in 1983, construction began in 1982 and was completed late in 1987 at a cost of approximately $60 million. The resulting highway is much less invasive than a standard Interstate would have been. It consists of 2.75 miles of standard Interstate, 1.75 miles of four-lane parkway, 2.5 miles of three-lane parkway, and three miles of two-lane parkway. The three-lane segments are configured with two lanes in the uphill direction to allow passing. It has a narrow or minimal median in several segments, and the thirty-foot clear zone on the roadside was reduced. The project also included significant improvements to recreational facilities, including the addition of a bike and pedestrian path and tourist parking areas. The project was twice a finalist for the American Society of Civil Engineers Outstanding Civil Engineering Achievement award.[20]

---

[19] E. H. B. Bartelink, personal communication with the author, May 26, 1988.
[20] New Hampshire State Department of Transportation, press packet (1988) author's files; "Ten Top Projects for 1988," *Civil Engineering* (July 1988): 68-70; "Top Projects of 1989," *Civil Engineering* (July 1989): 57.

All of these cases share strong similarities that speak to the larger developments of the highway program at this time. Each project involved construction through a sensitive environment, two rural, one urban. In all three cases, the original design according to conventional Interstate design protocols met with strong opposition from environmental or community interests. In each case, a community-oriented design team was formed to facilitate incorporation of criteria other than safety and technical efficiency of the highway. In each case, the resulting project incorporated features that increased the cost but that mitigated the environmental impact. And in each case, engineering design criteria had to be compromised in the interests of environmental concerns.

The Franconia Notch, Glenwood Canyon, and I-66 projects characterized a new approach to highway project design, one that actively sought to involve the community and incorporate environmental design considerations into the engineering process. But the effect of these changes was to stretch out the design process significantly. Incorporating environmental criteria into the design process in many cases involved a long series of interactions between engineers and community and environmental interests in order for the two parties to better understand the other's decision criteria. Furthermore, in each of the three cases, staunch opponents of any highway construction remained who refused to participate in a cooperative design process. To these opponents the remedies provided by NEPA and subsequent environmental policy and legislation allowed significant opportunity to delay through injunction and through calls for further environmental review.

The changes in process and in the participants in that process led to changes in design, as illustrated in each of these three thumbnail case studies. Arguably, the changes led to better engineering and moved away from "design by statute," that is, design parameters embedded in statutes. Stakeholders who value the various dimensions of a project differently may, through negotiation, reach a consensus on a design that balances safety, speed, cost, aesthetics, and environmental impact for a particular project.

But some of the underlying value conflicts cannot be resolved at the project level. Indeed, many remain unresolved in current debates about surface transportation policy. These issues include the role of automobiles and transit in urban transportation, the importance of preserving traditional urban forms, and the environmental impact of surface transportation.

"Finishing" the Interstate

As Interstate construction continued throughout the 1970s and 1980s, attention increasingly focused on two issues: resolving the unbuilt sections, either by completing them or deleting them from the system, and maintaining the Interstates that had already been built. By 1995, the Interstate was 99.9 percent complete, with only forty-six miles remaining unopened, forty of which were under construction and the balance of which were in engineering and right-of-way acquisition.[21]

---

[21] U.S. Federal Highway Administration, *Bulletin* (September 7, 1995).

THE POST-INTERSTATE ERA

As construction of the Interstate approached completion at the end of the 1980s, after thirty-five years of effort, the surface transportation policy community began to shift its attention to what should follow. The Interstate system was remarkable for the incredibly robust and consistent commitment that it won from Congress for decades. Whereas in other parts of the budget Congress is ever prone to earmark special projects benefiting their own districts, such behavior had been surprisingly rare in the Federal-Aid Highway Program. Notwithstanding all of the controversy of the freeway revolts and the environmental protection movement, the 41,000-mile system identified in the 1956 highway acts grew only 4 percent to 42,795 thirty-nine years later when it was complete.

Yet its status as a crash program raises problems as well. It was supposed to last only twelve years but took three times that long. In his classic treatise on military strategy, Henry Eccles comments on the danger of crash programs, or, as he calls them, "logistical snowballs." They tend to sap other programs of resources, thereby increasing overall costs and reducing effectiveness, leading to requirements for still more resources. In some ways, the Interstate program had just such an effect on American surface transportation.[22]

Discussions about redefining the federal role in highway policy began to emerge in the early 1980s under the rubric of President Ronald Reagan's "new federalism," which sought to reorder the allocation of responsibilities between the federal, state, and local governments. New federalism revived discussion of the Interstate program's role in national (i.e., interstate) and local transportation. A Congressional Budget Office (CBO) report of the time captures the tone of the debate.

> From a transportation viewpoint, roads that link activities in different states and contribute to interstate commerce are of prime national importance, while roads that serve local traffic needs or that link localities to the national network are of lesser national importance....
>
> [A] key feature of the Interstate program is its compound scope. It finances not only major intercity arteries that carry goods and people from state to state, but also major arteries within urban areas—roads serving commuting and other local needs.... Since the Interstate program began in 1956, it has financed both national and local roads, although this policy has been much debated both as it was being written into law and during the years the law has been applied.... [T]he dual national/local emphasis of the Interstate program, which has been its governing policy since 1956, has shifted complexion over time. First, in response to local concerns, a mechanism for withdrawing routes was devised, which implicitly recognized that certain, primarily local routes were dispensable. Now, in response to financial pressures, another policy change could

---

[22] Henry E. Eccles, *Military Concepts and Philosophy* (New Brunswick, N.J.: Rutgers University Press, 1965), chap. 7.

further channel Interstate resources into those key routes that are integral to a national, interconnected system of roads[23]

As the quotation makes clear, concentrating on routes of national significance would constitute a sharp break with the highway policy of the Interstate era and, in fact, the whole twentieth-century highway program.

Discussions of the national highway policy often overlook the historic dual emphasis of the highway program on national and local travel demand, which is grounded in the massive predominance of local over nonlocal or national travel. A few statistics make the point. Personal vehicle use is overwhelmingly local, whether measured in terms of trips or travel. Almost 90 percent of the nation's personal vehicle trips are less than 20 miles, comprising 55 percent of all personal vehicle travel (see Table 4-2). Ninety-four percent of personal vehicle trips are 30 miles or less, comprising 68 percent of personal vehicle travel. Trips over 30 miles comprise less than 5 percent of the trips and roughly one-third of all personal vehicle travel.

Table 4-2
Vehicle Trips and Travel vs. Trip Length, 1990

Trip Length	Trips %	Cum. %	Travel (VMT) %	Cum. %
≤5	61.5	61.5	15.5	15.5
6-10	16.4	77.9	16.6	32.1
11-15	7.7	85.6	13.2	45.3
16-20	4.2	89.8	9.9	55.2
21-30	3.8	93.6	12.6	67.8
>31	4.5	98.1	32.3	100.0
NR[1]	1.9	100.0	–	–
Total	100.0		100.0	

Source: U.S. Federal Highway Administration, *NPTS Data Book*, National Personal Transportation Survey (1990), 4-89, 5-36.
[1] Not Reported.

Truck travel is also predominantly local. Roughly half of all trucks are used for trips of 50 miles or less, and that comprises 63 percent of all truck travel (see Table 4-3). Trucks used for trips of 200 miles or less comprise 89 percent of the truck fleet and 86 percent of truck travel. Without understating the importance of long-distance travel, it is still fair to observe that, like politics, most transportation is local, to paraphrase Tip O'Neill.

---

[23] U.S. Congressional Budget Office, *The Interstate Highway System: Issues and Options* (Washington, D.C., June 1982), 18-20, 52.

Table 4-3
Truck and Truck Miles vs. Trip Length, 1997

Usual Trip Length[1]	Trucks (mill)	%	Cum. %	Truck Miles (bill)	%	Cum. %	Miles/truck (000)
Local (<50 mi)	52.8	72.5	72.5	657.8	63.0	63.0	12.5
Short-range (50-200 mi)	12.3	16.9	89.4	240.9	23.1	86.1	19.6
Long-range (> 200 mi)	3.8	5.2	94.6	119.4	11.4	97.5	31.4
Off-the-road	2.9	4.0	98.6	26.2	2.5	100.0	9.0
Not reported	1.0	1.4	100.0	(2)		100.0	NA
Total	72.8	100.0		1,044.2	100.0		14.3

Source: U.S. Bureau of the Census, *Vehicle Inventory and Use Survey* (Washington, D.C., 1997), table 2a.
[1] Length of trip in which truck is generally used.
[2] <0.05 percent.

The constitutional justification for the highway program and the appropriate federal role in a federal system is also a source of confusion. As in the CBO reported above, highway policy debates often identify interstate commerce as a basic rationale for the federal program. That basis is relatively new, however. Leaving aside the debates about federal support for "internal improvements" in the nineteenth century, the constitutional foundation of the highway program as it was enacted in 1916 was not interstate commerce but constitutional authority to establish a post office and post roads. The legislative history of the 1916 highway bill shows that Congress rested its authority for the program in that clause. Indeed, the early allocation formulas for the highway program used, as one factor, the mileage of post roads in a state.[24]

Clearly, the post roads justification for the federal program has fallen by the wayside. Support appears to be growing for justifying the program on the basis of the interstate commerce clause. Yet while policy makers seek constitutional guidance for sorting out the federal role, the Constitution provides broad discretion in defining a role, for actions not justified on the basis of interstate commerce might well be justified in the interests of general welfare. But the constitutional justification for the highway program was, and probably remains, a relatively minor issue.

The two reauthorizations of the 1990s—the Intermodal Surface Transportation Efficiency Act of 1991 (ISTEA) and the Transportation Equity for the 21st Century (TEA-21)—introduced important changes. While it is still too early to assess whether these two laws will fundamentally

---

[24] Jonathan L. Gifford, "An Analysis of the Federal Role in the Planning, Design and Deployment of Rural Roads, Toll Roads and Urban Freeways," Ph.D. diss., School of Engineering, University of California, Berkeley, 1983.

change the nature and scope of the highway program, the policy debate they have engendered has changed, and the participants in that policy debate have expanded significantly.[25]

The Intermodal Surface Transportation Efficiency Act of 1991 (ISTEA)

In his epic novel *The Man without Qualities*, Robert Musil describes the elaborate preparations, beginning in 1913, for the seventieth anniversary of the reign of the Austrian Emperor Franz Josef in 1918. The party planners are particularly intent upon outdoing a parallel German celebration of the thirtieth year of Kaiser Wilhelm's reign, also to occur in 1918. The irony is that neither of these empires will even exist when the planned celebrations are to occur because of the impending world war.[26] Preparations by the mainstream highway community for the reauthorization of the surface transportation act in 1991 resembled those for the emperor's anniversary, for their massive expenditure of energy on planning and coalition building was dramatically preempted by events totally beyond their expectations.

As Interstate construction stretched into the 1980s, Congressional earmarking for "demonstration projects," or to use the more colloquial term, "pork," began to grow. In 1987, President Reagan vetoed the highway reauthorization bill. While his veto was overturned (by a single vote), it was nonetheless the first time in history that a president had vetoed a highway authorization bill. The reason for his veto: it contained too many pork barrel projects. These pork barrel projects were a marked departure from the professional norms and the reliance on technical criteria that had been tried, tested, and strongly adhered to for decades. The pork reflected an erosion of Congress's willingness to give priority to the Interstate at the expense of other transportation priorities.[27]

Moreover, while the centerline mileage of the Interstate system had remained remarkably stable, funding provisions had been adjusted for some time to ensure a certain "fair sharing" of Interstate funds. Whereas funds were initially allocated according to the cost to complete the Interstate in a particular state, concern about some state shares being too small led to the institution of a 0.5 percent minimum in the 1960s. By 1987, when Reagan vetoed the bill, the number of explicit pork barrel projects had grown to 152.

The next reauthorization was due in 1991, and many in the highway community were concerned about the long-term viability of the federal program. One way to achieve this was to identify a well-defined system to replace the Interstate. Such a system could provide a focus

---

[25] "Intermodal Surface Transportation Efficiency Act of 1991," P.L. 102-240, 105 *Stat.* (18 December 1991): 1958. The "Transportation Equity Act for the 21st Century" (P.L. 105-178, 112 *Stat.* 107-509, 9 June 1998) authorizes the federal surface transportation programs for highways, highway safety, and transit for the 6-year period 1998-2003. The "TEA 21 Restoration Act" (Title IX, P.L. 105-206, 112 *Stat.* 834-68, 22 July 1998) provided technical corrections to the original law. This book refers to the combined effect of these two laws as TEA-21.
[26] Robert Musil, *The Man without Qualities*, translated by Sophie Wilkins and Burton Pike (1930-43; New York: Knopf, 1995); "Robert (Edler von) Musil, 1880-1942," in *Contemporary Authors Online* (The Gale Group, 2000).
[27] The facts of the legislative history in this section draw heavily on Jonathan L. Gifford, Thomas A. Horan, and Louise G. White, "Dynamics of Policy Change: Reflections on 1991 Federal Transportation Legislation," *Transportation Research Record* 1466 (1994): 8-13.

much like the Interstate had, and counter the political forces favoring pork. The mainstream highway community, led by the American Association of State Highway and Transportation Officials (AASHTO) and the Highway Users Federal for Safety and Mobility (HUFSAM)[28] mobilized a broad-based initiative called "Transportation 2020" with the avowed purpose of assessing the strengths and weaknesses of American highway and surface transportation policy and developing a policy drawing from this coalition that would receive widespread, if not universal, support on Capitol Hill at the time it was presented. Transportation 2020 sponsored hearings in each state and the creation of technical advisory groups to solicit and synthesize the views of all interested parties.

AASHTO also sponsored a history and review of experience for lessons to be learned from the construction of the Interstate program. This review was to be, in the words of Frank Turner, one of the project's advisors and the "godfather" of American highway policy in the postwar era, a "management audit," an objective assessment of the mistakes that were made and the lessons that could be learned from the Interstate highway program. This review conducted a national survey and over one hundred interviews with major figures then on the national scene as well as state highway officials.[29]

The rationale behind Transportation 2020 might be paraphrased as follows: "if we can just get everyone participating in the process, get everyone's ideas out on the table, we will be able to forge a policy proposal that will receive unanimous support." Attention and energy were therefore directed at assembling that group around the table and eliciting their views on the appropriate course and direction of American surface transportation policy.

Key issues for the Transportation 2020 were the need for a system of highways of national significance, and the need to simplify the categorical structure of the program. Separate funding categories had proliferated over the years, and many state departments of transportation chafed at restrictions on using funds allocated to one category where state needs might not be acute to another where they were.[30]

In some respects, this reflected a longing for the days when the executive branch would introduce a highway bill and forty-eight state highway department heads would testify in favor of it before the congressional committees. That is, it was a time of consensus and unity about the objectives of American surface transportation policy.

Meanwhile, the administration initiated a parallel set of initiatives under the rubric of a strategic plan by Secretary of Transportation Samuel Skinner. The administration was partially con-

---

[28] HUFSAM later changed its name to the American Highway Users Alliance, which sponsored the fortieth anniversary celebration on the Ellipse in 1996.
[29] The author participated in this study, American Association of State Highway and Transportation Officials, *The States and the Interstates*.
[30] American Association of State Highway and Transportation Officials, Task Force on a Consensus Transportation Program, *New Transportation Concepts for a New Century: AASHTO Recommendations on the Direction of the Future Federal Surface Transportation Program and for a National Transportation Policy*, final ed. (Washington, D.C., October 1989).

cerned with not increasing the federal budget deficit. Early in 1991, it unveiled its post-Interstate proposal after the president introduced it in his State of the Union address. Like the proposal from Transportation 2020, the administration focused on the need for a nationwide system of highways to replace the Interstate and on the need for greater flexibility in the use of funds.[31]

A third set of actors with an interest in reauthorization was the members of Congress who had grown fond of their pork barrel projects. Consistent with its more local orientation, the House of Representatives traditionally initiated action on the transportation bill, which fell under the jurisdiction of the Committee on Public Works and Transportation, at that time newly chaired by Representative Robert Roe (D-N.J.), who had campaigned for chair on promises of new projects. To support such projects, Roe was engaged in an effort to raise the gas tax by 5 cents. He linked success for his "Nickel for America" proposal directly to $6.8 billion in what he called "projects of national significance" (that is, pork).

The "Nickel for America" posed significant strategic problems for the traditional highway community. Roe tied it directly to pork barrel projects, which they opposed in favor of a technically oriented selection process controlled by the federal and state departments of transportation. Moreover, the president opposed tax increases, as did an important member of the highway coalition, the trucking industry. Roe eventually had to drop his plans, but his efforts to push through the tax increase splintered the leadership in the House.

The issue of identifying highways of national significance was particularly divisive. AASHTO had taken the lead in attempting to work with the states to identify such a system, but eventually withdrew due to a high level of controversy among the states on how national significance was to be defined. Initial attempts to identify the system on the basis of functional classifications broke down when states indicated that some of their functional classifications needed to be revised to reflect current conditions.[32] Functional classification was supposed to be a purely technical definition of a road's function. But using functional classification as a means of determining funding eligibility would have badly undermined its utility as a technical criterion. Congress, meanwhile, was not interested in a network that would be determined solely by the federal and state departments of transportation. It wanted a chance to review and approve a map and, of course, to add to that map if it so desired.

Finally, a new coalition had arrived on the scene that was interested in significant reforms to the highway program, which members believed encouraged single-occupant vehicle travel, air pollution, and suburban sprawl. A group of urban planners and environmental groups had formed during the 10-year effort to revise the Clean Air Act, which culminated in the Clean Air Act Amendments of 1990 (CAAA).[33] Recognizing that they could significantly enhance the prospects for air quality improvements by leveraging them with the substantial resources of

---

[31] U.S. Department of Transportation, *Moving America: New Directions, New Opportunities* (February 1990).
[32] Author interviews with Kevin Heanue, Federal Highway Administration, and Janet Oakley, National Association of Regional Councils (November, 1990).
[33] "Clean Air Act Amendments of 1990," P.L. 101-549, *Stat.* 104 (15 November, 1990): 2399.

the highway program, the group reassembled as the Surface Transportation Policy Project (STPP). STPP's sponsors included the National Trust for Historic Preservation, the American Institute of Architects, and some dozen other civic and environmental groups that largely opposed traditional highway program.

The Surface Transportation Policy Project seized on the opportunity presented by the surface transportation reauthorization to move national clean air policy beyond a purely regulatory approach by using highway funds as incentives for compliance or, more accurately, by withholding highway funds as a penalty for not attaining national air quality standards. They also sought to dilute the authority of state departments of transportation in highway construction because they believed that new highways led to increases in air pollution by encouraging auto-dependent lifestyles and development patterns. Their interest in transportation was a natural extension of their interest in the environment:

> We knew early on that clean air was going to be driving a lot of where the [Senate] committee was going.... The transportation debate has been so overwhelmingly dominated by the highway community for so many years. The nature of what the committee did on clean air should have been a signal to the highway community.[34]

STPP also had close ties to the Senate Committee on Environment and Public Works, which had jurisdiction over both the Clean Air Act and the surface transportation program. The subcommittee with jurisdiction over reauthorization was chaired by Daniel Patrick Moynihan (D-N.Y.), who shared a strong commitment to significantly reforming the highway program. Moynihan had long had an interest in urban issues and public works, and saw 1991 as an opportunity to rethink the highway. He had written an article in 1960 called "New Roads and Urban Chaos" that was critical of the new Interstate program. He had also been instrumental in the creation of the Pennsylvania Avenue Development Corporation, which starting in the 1960s had revitalized the derelict boulevard connecting the White House and Congress.[35]

Moynihan had an unusually free hand as subcommittee chair with the surface transportation reauthorization. The chair of the full committee, Quentin Burdick (D-N.D.), was aging; he died in office a year later. Burdick had been a fairly weak chair since his reelection in 1988. Moynihan also forged an alliance with other powerful senators on the committee. The ranking minority member of the subcommittee, Senator Steve Symms (R-Idaho), supported devolution of authority to states and localities. Senator Frank R. Lautenberg (D-N.J.) was also sympathetic to Moynihan's views, as was longtime environmental advocate Senate John H. Chafee (R-R.I.), who was the ranking minority member of the full committee.

Perhaps the most dramatic event of the 1991 reauthorization was the release by Senate Committee on Environment and Public Works of its markup of the bill. Traditionally, the House has played the dominant role in the marking up of highway legislation, consistent with its more lo-

---

[34] Janet Oakley, National Association of Regional Councils, quoted in *Congressional Quarterly Weekly Report* (May 25, 1991), 1367.
[35] Daniel P. Moynihan, "New Roads and Urban Chaos," *The Reporter*, April 14, 1960, 13-20.

calized focus. Little or no substantive change was expected from the Senate. Imagine the surprise when Moynihan's committee unveiled a surface transportation bill that specified fundamental changes in American highway policy. Perhaps most striking was the absence from the bill of any designated national highway system—a cornerstone of the Federal-Aid Highway Program since 1921. Contemporary reporting of the event reflects the surprise of the highway community.

> In a significant victory for a coalition of environmentalists and urban planners, the Senate Environment and Public Works Committee on May 22 approved a five-year surface transportation bill that would radically alter federal highway policy.... Most remarkable about the Senate bill is that it was crafted with the interest of environmentalists and urban planners in mind, rather than those of the traditional highway lobbyists who have typically left their imprint in such reauthorizations. The so-called road gang of highway lobbyists was focused on the House Public Works Committee, which traditionally has taken the lead in introducing such bills, when the Senate bill was unveiled. The group includes the American Trucking Associations, state transportation officials, motor vehicle manufacturers and the Highway Users Federation.[36]

The highway community, which had been bickering over details of a more conventional reauthorization, was caught flat-footed by the introduction of Senator Moynihan's bill. The efforts of Moynihan and STPP, who had taken pains to keep a low profile, especially vis-à-vis members of the traditional highway lobby, had eclipsed the careful efforts of Transportation 2020.

The bill that finally became law was the Intermodal Surface Transportation Act of 1991, or ISTEA. It passed on November 27, and the president signed it into law on December 18. During the previous summer the Senate had moved quickly and passed its bill on June 19. The House bill had a much more troubled legislative course. Roe's Nickel for America proposal encountered opposition. It was pulled from the floor on August 1 and formally abandoned on September 18. On October 10, a revised House bill was introduced, which the committee approved on October 15. The bill was passed by the full House on October 23. A 20-day conference then worked out the differences with the Senate.

One of the keywords in ISTEA was "intermodal," a new term in the transportation lexicon. The word represented a new approach to transportation that was not solely focused on the single mode of highways. In the passenger domain, this reflected a desire to include non-highway modes. In the freight domain, it reflected a desire to attend to connections between modes, such as ports and rail yards.

ISTEA contained a number of important changes, including several that had been contained in the Senate bill. It provided increased flexibility to state, local, and regional planning entities, but also placed them under new obligations requiring openness to public dialogue and input. It also provided greater flexibility for funding non-highway modes, including carpools and

---

[36] "Overhaul of Highway System Approved," *Congressional Quarterly Weekly Report* (May 25, 1991), 1366.

vanpools, transit, commuter rail, and municipal bikeways. Yet it did not mandate much reallocation of spending. Of the $151 billion authorized, $110 billion could be spent on any mode.[37]

ISTEA also required states to develop and implement six management systems in cooperation with metropolitan planning organizations (MPOs): pavement on federal-aid highways, bridges on and off federal-aid highways, safety, traffic congestion, public transportation facilities and equipment, and intermodal transportation facilities and systems. States objected to this mandate, however, and the requirement was relaxed to an eligible activity in 1995.

ISTEA also authorized $6 billion for a new funding category, congestion mitigation and air quality (CMAQ) for use in areas not in compliance with air quality standards. It increased funding for metropolitan areas and strengthened the role of MPOs in the selection of projects that receive federal funds, required the consideration of air quality and energy impacts of transportation investments, and introduced new requirements for public participation in decision making. On the issue of defining a system of highways of national significance, the desire of Congress to approve a map prevailed, and the bill stipulated that U.S. DOT would present a map of a new national highway system, a successor to the Interstate, for congressional approval by September 30, 1995.

Another important provision of ISTEA was the creation of a new category of spending known as "enhancements," funded with $400 million. The idea behind enhancements was that transportation funding should be available not only for transportation facilities and services, but also for enhancements to the environment in which they operated. Enhancements were, depending on one's point of view, an invaluable tool for compensating communities for the impacts of transportation projects, or a shameless rip-off of the transportation taxpayer. Enhancement funds were used to support bicycle paths, rehabilitate historic buildings, and acquire scenic easements along highways.[38]

National Highway System Designation Act of 1995

The next major round of transportation legislation revolved around the designation of the national highway system (NHS) in 1995. ISTEA had required Congress to designate the NHS by September 30, 1995, as a condition for states to receive $6.5 billion in annual federal-aid funds.

The development of the system that FHWA submitted to Congress engendered some significant conflict between the federal and state interests. What FHWA sought was to develop a system that kept the best features of the Interstate system—a limited, interconnected network that would be the target of significant investment—and the NHS as submitted to Congress fell

---

[37] Jonathan L. Gifford, William J. Mallett, and Scott W. Talkington, "Implementing Intermodal Surface Transportation Act of 1991: Issues and Early Field Data," *Transportation Research Record*, no. 1466 (1994): 14-22.
[38] Joseph R. DiStefano and Matthew Raimi, *Five Years of Progress: 110 Communities Where ISTEA Is Making a Difference*, <http://www.Transact.Org/5yrs/Index.Htm> (Washington, D.C.: Surface Transportation Policy Project, 1996).

far short of that. It contained many miles of residential streets in urban areas unlikely to ever be the target of significant investment.

States pushed hard to designate as many miles of NHS as they could get in the belief that it would some day be part of a formula for funding allocation. The Office of Management and Budget (OMB), however, was concerned about the potential budgetary consequences of such a large system and sought to reduce the total mileage, which created conflicts with a number of states. After a series of negotiations and extensive meetings, a resolution was reached and submitted to Congress in 1993.[39]

In Congress, the NHS designation itself was not particularly controversial. Its status as a bill that had to pass tempted members to add other provisions, which led to controversy, delay, and a protracted conference of almost two months. The president signed the bill into law on November 28, 1995, over the vociferous opposition of highway safety advocates.[40]

The law designated a 161,000-mile system. The only real controversy over the NHS itself lay in whether the law would allow U.S. DOT to make changes to the routes at the request of a state without congressional approval, which it ultimately did.

The controversial aspects of the law related to a number of regulatory mandates. The NHS act repealed the federal 55-mph speed limit, enacted in 1973 in response to the oil embargoes but retained because of its presumed safety benefits. It also repealed a requirement for states to require motorcycle helmets, to convert to the metric system, and to develop and implement the six management systems required by ISTEA.[41]

ISTEA also continued the expansion of pork barrel projects. Compared with the 152 projects enumerated in the 1987 bill vetoed by President Reagan, ISTEA contained 538 projects valued at $6.2 billion, approximately 4 percent of the bill's $156 billion total.[42]

Transportation Equity Act for the 21st Century (TEA-21)

ISTEA funded the surface transportation program through September 30, 1997, so reauthorization initiatives began well ahead of time in 1995 and 1996. The fortieth anniversary of the Interstate system in June 1996 on the Ellipse in front of the White House was one of the events staged to build public awareness and support. This time the coalition was called "Keep America Moving," and it involved approximately forty trade associations, including the American Road and Transportation Builders Association (ARTBA), the American Trucking Associations (ATA), the National Association of Manufacturers, and the Chamber of Commerce.[43]

---

[39] Kevin Heanue, telephone interview (March 5, 2001).
[40] "National Highway System Designation Act of 1995," P.L. 104-59, *Stat.* 109 (28 November 1995): 568.
[41] "National Highway System Bill Clears," *Congressional Quarterly Almanac* 51 (1995): 3-60–3-65.
[42] Ben Wildavsky, "Pigging Out," *National Journal* 29, no. 16 (April 19, 1997): 754.
[43] Peter H. Stone, "From the K Street Corridor," *National Journal* 29, no. 1 (January 4, 1997): 28.

A telling episode in the way Washington deals with new constituencies was STPP executive director Hank Dittmar's appearance on a live broadcast on reauthorization in the summer of 1996. On the panel with him were members of the traditional highway coalition, including Jack Schenendorf, a chief staff member from the House committee, Jane Garvey, deputy administrator of FHWA, Tom Larson, former FHWA administrator, the former Connecticut transportation secretary, and a city transit manager. Dittmar spoke passionately of the urgency of "getting the program level up" for reauthorization. STPP had been brought into the fold. Other environmental groups signed on to support reauthorization the following summer.[44]

As reauthorization played out in 1997, one of the central debates on the bill had to do with pork. House Public Works and Infrastructure Committee chairman Bud Shuster (R-PA) was a shameless advocate of member projects. "Angels in heaven do not decide where highways and transit systems are going to be built," he said. "It's a political process." House Budget Committee chairman John Kasich (R-OH) was staunchly opposed to the projects as budget busters and out of tune with the new agenda in Congress that had been ushered in with the Republican victory of 1994.

Shuster won hands down, although other matters delayed full reauthorization until 1998. The final bill included a record 1,850 highway projects worth $9.35 billion, another $3 billion in mass transit projects, and $3 billion in bus projects.[45]

Another key issue was the overall spending level. Shuster had repeatedly sought to take the Highway Trust Fund "off budget" to prevent positive balances in the trust fund from offsetting spending elsewhere. In TEA-21 he succeeded through the enactment of a complex set of "firewalls" that set spending levels based on the revenues to the trust fund in the prior year. While Shuster trumped the new Republican regime's budget guidelines, another of their rules cost him his chairmanship in 2001, when six-year term limits took effect. In a surprising move, he resigned his seat and endorsed his son to succeed him.[46]

* * *

American highway and urban transportation planning and development policy for the last century is deeply rooted in principles of rational order, subject to objective, scientific specification and validation. As in many fields of inquiry, support for that rational basis for policy making has eroded considerably.

Interests and stakeholders whose decision-making criteria have traditionally fallen outside the boundaries of that rational basis have facilitated this erosion. As in the cases of Franconia

---

[44] "The Reauthorization of ISTEA," live broadcast (Raleigh: North Carolina State University, Center for Transportation and the Environment, August 14, 1996); Margaret Kriz, "Road Warriors," *National Journal* 29, no. 26 (June 28, 1997): 1327.
[45] "Transportation Law Benefits Those Who Held the Purse Strings," *Congressional Quarterly Almanac* 54 (1998): 24-3, 24-5.
[46] "Politics," *National Journal's Congress Daily,* January 4, 2001; "Politics," *National Journal's Congress Daily,* February 20, 2001.

Notch, Glenwood Canyon, and I-66, competing interests have forced their way into the design and location process through broader participatory planning and decision making. This vision of community involvement and broad-based decision making are the foundation of the public participation requirements set forth in ISTEA. One of the consequences of the increased role of public participation in decision making has been the attenuation of the project and planning cycle. Projects can dwell for more than a decade in the planning stage while public participation and legal issues are resolved.

Another set of interests that has eroded the rational basis of transportation planning and decision making is—call it what you will—projects of national significance, projects of local significance, or just plain pork. For the last two decades, reauthorization legislation in Congress has raised the ante of member-specified projects, and there is little prospect for reversing that trend. The next two chapters examine in more detail how transportation planning has evolved, and the challenges it now faces.

5

Transportation Planning Methods

One of the central and abiding features of the Federal-Aid Highway Program has been its emphasis on planning and engineering. Decisions based on political or spur-of-the-moment considerations were not acceptable. The 1921 act required states to designate a system of highways, and only projects within that system were eligible for federal reimbursement. Project planning required the development of "plans, specifications and estimates" (PS&E) that were subject to the approval of Bureau of Public Roads (BPR) engineers before a project could go forward with the expectation of partial payment from the federal-aid program.

As the program developed over the following decades, planning requirements at both the system and project level developed as well. The advent of higher speeds and traffic congestion in the 1930s brought planning and engineering requirements to accommodate high speeds with geometric design and high traffic volume with appropriately scaled and access-controlled facilities.

From the start, a particular concern was the longevity of improvements, that is, ensuring that improvements financed with federal aid had lasting value. Early in the program, this emphasis motivated requirements for location studies, proper drainage, and the use of suitable construction materials.

The lesson the BPR learned in the 1930s was that changes in demand could reduce the value of highway improvements, even when they were properly constructed in good locations. Increased speeds had rendered some improved highways unsafe. Increased traffic volumes had led to congestion on other highways. The solution was to anticipate changes in demand, just as any good engineer anticipates variations in storm flows when sizing a culvert. But anticipating traffic speeds and volumes turned out to be a much bigger problem than estimating storm flows.

80 FLEXIBLE URBAN TRANSPORTATION

Longevity of improvements is largely an engineering value; it emphasizes the sanctity of engineering materials. While superficially appealing, its realization in the planning and design procedures under the Interstate program was to have profound—and often troubling—consequences for the urban Interstates. The principal difficulty was that while engineering materials may last for twenty years or longer, neither the demand for nor the supply of highways and other urban transportation developed according to prediction. Indeed, a central tenet of this book is that assuming predictability—and requiring predictions as a condition of federal aid—has been a fundamental weakness of the federal aid program that has had harmful—and unrecognized—consequences.

The quest for physical and functional longevity of highway improvements is at bottom a question of methodology: what methods and procedures should be used to determine the location and design of facilities so as to ensure their physical and functional longevity? This has been a central question for highway planning ever since.

The two previous chapters described the evolution of the Federal-Aid Highway Program and the Interstate program from the former's inception in the late nineteenth century to the latter's completion a century later. This chapter examines methods and procedures of highway and transportation planning. It begins with a discussion of the role of analysis in planning and decision making. It then turns to generic analytical techniques and approaches to planning and decision making. The next chapter examines the application of these methods to transportation planning at the metropolitan, regional, and national level.

THE INSTITUTIONAL CONTEXT OF TRANSPORTATION PLANNING

Transportation planning is the first step in decision making about the provision of transportation facilities and their operation. Authority over road and highway planning has traditionally resided primarily in the public sector. Private interests have participated either as stakeholders in the political and administrative process or as consultants to public agencies. In a few cases, private sector organizations have taken a leading role, as in the case of the Regional Plan Association in the New York metropolitan region.

American transportation planning is inextricably bound up with the way American society goes about setting goals and objectives. American social goals and objectives are often unclear; they are complex and often conflicting. American society enjoys unparalleled market freedoms and democratic liberties, and these conditions reflect basic beliefs of the American people about what constitutes a good society.[1]

These economic and political freedoms are rooted in a distinctly American ideology that dates back to Hamilton, Madison, and Jefferson and the founding of the nation. What Alexis de

---
[1] Portions of this section are adapted from Jonathan L. Gifford and Thomas A. Horan, "Transportation and the Environment," ms., 1994.

Tocqueville called "American exceptionalism"[2] runs deep in America and includes four basic traits:

- Individualism (personal and private initiative);
- Anti-statism (skepticism of government authority);
- Populism (wisdom and power of the common person); and
- Egalitarianism (equality of opportunity).[3]

This political culture stems in part from America's establishment as a revolutionary state in defiance of the power and authority of the English crown. In American political culture, government does not set goals and objectives for the American people. The people set goals through their participation in politics and collective action, in elections, through special interest lobbies, in legislation, and through other civic institutions.[4]

The principal mechanism for determining public policy is voting, either in elections or through one's behavior as a consumer. Voting in elections has the advantage of affording the opportunity for participation relatively equally, regardless of social or economic station. Yet it has a number of disadvantages. Narrow special interest groups often dominate political participation. Also, political processes give little indication or expression of willingness to pay for improved services. When decisions derive from a political and administrative process, it is difficult, for example, for a consumer to express directly a willingness to pay marginally more for better roads or for a freight hauler to express willingness to pay for improved travel time reliability.

Yet voting is also subject to a paradox. In the "paradox of voting," under common conditions the outcomes of a fair voting process can depend as much on the sequence of votes as on the preferences of the voters. As a result, those who control the sequence of votes can exert tremendous influence on voting outcomes.[5]

For example, suppose a referendum were held to choose among three alternative policies: a $3 per gallon gas tax, restricted spending and increased environmental standards for highway projects, or a continued flow of funding for highways and unchanged environmental standards. After much debate, all voters fall into one of three categories. Greens, who comprise 10 percent of the population, fear that continued reliance on the automobile will lead to global warming and environmental devastation. They seek a radical change in public behavior, which can only be brought about by a high gas tax. Their preferred policy is the gas tax; their second choice is limited funding and greater environmental standards.

---

[2] Alexis de Tocqueville, *Democracy in America*, 1833 (New York: Knopf, 1948).
[3] Seymour Martin Lipset, "American Exceptionalism Reaffirmed," in *Is America Different?* ed. B. Shafer (New York: Oxford University Press, 1991), 1-45.
[4] Robert D. Putnam, *Bowling Alone: The Collapse and Revival of American Community* (New York: Simon & Schuster, 2000).
[5] Kenneth Arrow, *Social Choice and Individual Values*, 2nd ed. (New Haven, Conn.: Yale University Press, 1963).

Pragmatists, who make up 45 percent of the population, want to discourage more driving, but do not want to bear the economic cost of a high gas tax. Their preferred policy is restricted funding for highways and increased environmental standards; their second choice is continued funding for highways and unchanged environmental standards.

Finally, constitutionalists, who comprise 45 percent of the population, prefer continued highway funding and unchanged environmental standards. But if action is to be taken against increased driving, they prefer the direct approach of higher gas taxes rather than the contentious indirect route of restricted highway funding and increased environmental standards.

Table 5-1 summarizes the groups' preferences and the results of voting under three different sequences of voting questions. Agenda A offers the gas tax proposal against
continued funding. Greens and pragmatists both prefer the gas tax to continued funding, so the gas tax wins 55 percent to 45 percent. The next round offers voters a
choice between the gas tax and continued funding, and continued funding wins 90 percent to 10 percent because both pragmatists and constitutionalists prefer continued funding to the gas tax. Agendas B and C offer different sequences of paired issues, leading to a final win for reduced funding and the gas tax.[6]

Table 5-1
Illustration of the Paradox of Voting

Hypothetical Groups	1st Choice	2nd Choice	3rd Choice	% of Voters
Greens	Gas tax	Reduced funding	Continued funding	10
Pragmatists	Reduced funding	Continued funding	Gas tax	45
Constitutionalists	Continued funding	Gas tax	Reduced funding	45

Agenda A (Result: Continued funding)
Round 1: Gas tax vs. Reduced funding   Gas tax wins 55% to 45%
Round 2: Gas tax vs. Continued funding   Continued funding wins 90% to 10%

Agenda B (Result: Reduced funding)
Round 1: Gas tax vs. Continued funding   Continued funding wins 90% to 10%
Round 2: Continued funding vs. Reduced funding   Reduced funding wins 55% to 45%

Agenda C (Result: Gas tax)
Round 1: Continued funding vs. Reduced funding   Reduced funding wins 55% to 45%
Round 2: Reduced funding vs. Gas tax   Gas tax wins 55% to 45%

Source: Adapted from David L. Weimer and Aidan R. Vining, *Policy Analysis: Concepts and Practice*, 2nd ed. (Englewood Cliffs, N.J.: Prentice Hall, 1992), 114-15.

---

[6] Adapted from David Leo Weimer and Aidan R. Vining, *Policy Analysis: Concepts and Practice*, 2nd ed. (Englewood Cliffs, N.J.: Prentice Hall, 1992), 114-15.

The public also expresses its transportation preferences through its behavior, within the range of facilities and services that are currently available. Hence observations of behavior are a good, if incomplete, indicator of preferences. But behavior says little about demand for services that are not currently available. Behavior is also subject to the current distribution of income and wealth—in most cases the poor necessarily have much less ability to pay than the rich. Yet willingness to pay is an important indicator of transportation system performance. If the system provides services for which users are unwilling to pay, that is a strong indicator that performance of the system is not up to par. For example, for all of its difficulties and unpleasantness, traffic congestion is a strong indicator of consumers' willingness to pay for the services provided by the existing transportation system, both with their time and with their expenditures for automobiles, within the range of choices and at the prices currently available to them.

Both political participation and behavior provide valuable information about how society values the inputs to and outputs of transportation facilities and services. Government attempts to encourage policies that lead to efficient, fair, and sustainable public systems, including transportation. But there are usually competing goals and tradeoffs. Environmental preservation may increase construction costs, and higher gasoline taxes may limit the mobility of the poor.

Any examination of the goals and objectives of the nation's transportation system must recognize the limitations of uniform, broad-brush, top-down approaches and seek guidance from people's preferences, demands, and behavior. Innovations, plans, or system developments that do not accommodate this belief system will most likely not succeed. Transportation policies that go against these basic traits may be difficult to implement.

THE ROLE OF ANALYSIS IN TRANSPORTATION PLANNING

Transportation planning covers a broad spectrum of decisions ranging from setting goals and objectives for national transportation systems at one extreme to the prioritization of individual projects in a local capital program at the other. As such, it engages two types of decision making. The first is the selection of which goals should be primary and which should be secondary, and the second is operationalizing agreed upon goals. Cook calls these "instrumental" and "constitutive" rationality, respectively.[7]

Traditional public administration held that selecting goals was the province of politics, while operationalizing agreed upon goals was the province of administration. Transportation planning would be much simpler if politics was the forum for resolving disputes over values, and administration the forum for implementing such decisions. Planning would fulfill an essentially technical function of operationalizing clear, unambiguous political decisions.

---

[7] Brian J. Cook, *Bureaucracy and Self-Government: Reconsidering the Role of Public Administration in American Politics* (Baltimore: Johns Hopkins University Press, 1996).

Bureaucracies are traditionally quite comfortable with the idea of improving procedures and best practices.

But the notion that bureaucracies objectively operationalize policy that is developed by legislatures and elected politicians has lost favor, and this is certainly the case in transportation. The boundary between politics and administration is soft. Bureaucratic "administration" and implementation of programs contains considerable policy content. Laws and legislative history are often ambiguous, which can allow multiple competing interests to claim victory in a dispute, deferring further resolution to implementing agencies and the courts. Equally, the technical content of competing policy proposals can be so complex that the bureaucrats in charge of developing legislative options significantly influence which options are debated.

Objective expertise and analysis therefore play an important part in transportation planning, but only a part. Analysis competes with other perspectives, such as the views of affected stakeholders and those of policy advocates.

As a result, the degree of consensus about the goals of transportation policy and the means for achieving those goals powerfully influences the nature of debates on the subject. When there is consensus about goals and how to achieve them, policy may be based largely on analysis. When there is disagreement about goals and how to achieve them, policy may be based on discussion or political processes (see Figure 5-1).

Figure 5-1
Framework for Analyzing Goals and Means

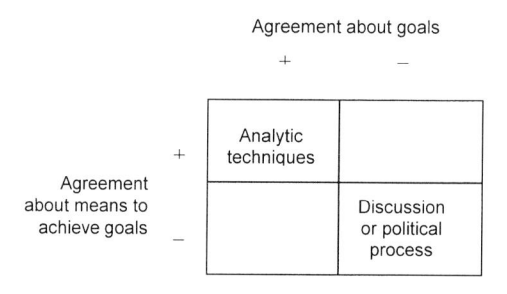

	Agreement about goals +	Agreement about goals −
Agreement about means to achieve goals +	Analytic techniques	
Agreement about means to achieve goals −		Discussion or political process

Source: Adapted from James D. Thompson, *Organizations in Action:* (New York: McGraw-Hill, 1967).

Analytic techniques are framed and debated in a manner that is distinctly different than discussion or political processes. One framework for understanding these competing perspectives identifies three modes of "discourse": analytical, critical, and persuasive.

Analytical discourse derives from the scientific model and is based on objective standards of evidence and rules for drawing inferences. Critical discourse focuses on the underlying value systems and power structures that inform and influence the policy process, such as whose values are dominant in decision making, and how power, authority, and resources are distributed. Persuasive discourse focuses on advocacy of a particular perspective or body of interests.[8]

Transportation planning under the federal-aid and other grant programs is firmly rooted in analytical discourse. It relies heavily on the values of verifiable scientific technique and procedure. But the other modes of discourse are very visible in today's transportation policy environment. Critical discourse is much more prevalent in the discussion of the social implications of transportation policy, such as how environmental and economic impacts are distributed across society over time and generations, and what that implies for general welfare. Persuasive discourse is most prevalent in the advocacy community, which advocates the environmental, community, or social agendas of interest groups.

In reviewing policy in the transportation sector it is important to recognize the presence—and the legitimacy—of each form of discourse. Analytical discourse and the scientific paradigm are most widely accepted in the engineering community, yet they are based on value judgments about the legitimacy of mathematically framed models and techniques. It is not necessary to dismiss entirely the legitimacy of analytical discourse to appreciate the legitimacy of critical and persuasive discourse. Much of the dispute over transportation policy derives not from divisions within each mode of discourse about the nature of its subject matter, but rather from tensions among the three. In understanding transportation policy, recognition of the presence of these modes of discourse and an appreciation of their legitimacy with different groups can provide helpful insights into the nature and unfolding of the policy process.

DECISION-MAKING CONCEPTS

Planning and decision making for transportation facilities and services encompass a broad range of approaches, from those based squarely in the analytical camp to those based in criticism and advocacy. The following sections set forth the basic tenets of a range of approaches. The first section examines approaches derived from indicative planning, the next turns to those based in microeconomics, and the last turns to those based in participation by stakeholders.

Indicative Planning

Indicative planning means planning and decision making that relies on "indicators" of adequacy for a particular condition or situation. For example, highway planning relies on an

---
[8] Louise G. White, "Policy Analysis as Discourse," *Journal of Policy Analysis and Management* 13, no. 3 (Summer 1994): 506-25.

assessment of "level of service" (LOS), which can range in value from "A" (very good) to "F" (very poor). Air quality planning relies on an area's conformity with a set of National Ambient Air Quality Standards (NAAQS), which establish "safe" levels of six pollutants.[9]

Indicative planning therefore relies on two stages of analysis. The first is the establishment of the indicators themselves and the determination of their acceptable values. The second is the assessment of conformity with those values.

The principal appeal of indicative planning is its administrative efficiency. It fits well with a centralized bureaucratic form of decision making and administration. The two-stage nature of the analysis lends itself to the concentration of expertise and decision-making authority in a central body that establishes the indicators and their acceptable values. Field offices can then determine conformity with those values and forward resource and action requests to the central office, which can "roll up" regional and national summaries of field conditions in a simple and efficient manner, based on uniform definitions and classifications.

Indicative planning breaks down, however, when experts or stakeholders disagree about the selection of indicators or the determination of acceptable levels of indicators. If the science that determines the indicators and their acceptable values is not air tight, stakeholders who are advantaged or disadvantaged by a particular indicator or threshold level might seek to undermine it. Communities that find themselves out of conformity with the NAAQS, for example, might appeal to Congress for relief on the grounds that the science is bogus.

A further concern about indicative planning is that it only crudely incorporates the costs and benefits of achieving conformity with indicator values. Both the cost and benefit of conforming to a particular indicator may vary widely depending on local conditions. The benefits of cleaner air, for example, might be much greater in a community with a high concentration of the elderly. The costs of conformity might be much greater in an area with a high concentration of extractive industries. The cost of remaining within a specified highway design criterion might be much higher in an area with high land values or mountainous terrain.

Indicative planning can accommodate such local conditions, but only crudely. It can stratify the acceptable indicator values to accommodate important local variations. The NAAQS might be stratified to reflect the age distribution of the population, for example, specifying more stringent standards in communities with a higher concentration of the elderly. Alternatively, the standard might allow exceptions for situations where costs of compliance are excessive. The definition of excessive, however, then comes into play. When the interests of all parties are closely aligned, and when all interests share the same set of basic values, resolving such disputes can be consensual. But when interests are at cross-purposes, or when values conflict, resolving such disputes can be very difficult.

---

[9] The six criteria pollutants are total suspended particulates (TSPs), sulfur dioxide (SO_2), carbon monoxide (CO), nitrogen dioxide (NO_2), ozone (O_3), and hydrocarbons (HC). Congress authorized the regulation of these pollutants in the "Clean Air Act Amendments of 1970," P.L. 91-604, 84 *Stat.* 1676-713 (31 December 1970).

Margin of Safety Analysis

One variant of indicative planning is so-called margin of safety analysis, whereby a relatively arbitrary margin of safety is established at, say 90 or 99 or 99.9 or 99.99 percent. The corresponding level of risk is 10^{-1}, 10^{-2}, 10^{-3}, or 10^{-4}. Margins of safety in many domains specify risk at 10^{-6}, or one-in-a-million probability. The margin of safety then provides the criterion for decision making.

Margin of safety analysis is widely used in the areas of human and environmental toxicology. Based on scientific evidence and expert judgment, regulators establish threshold levels of a pollutant above which the margin of safety is breached. The objective of policy is to maintain or achieve concentrations of a pollutant below those thresholds. Hence, indicators drive policy.

Regulation based on margin of safety analysis became increasingly important with the spate of environmental regulations in late 1960s and the 1970s. A primary objective of national policy was the elimination of the most visible and most clearly noxious pollutants from the everyday environment. The primary motivation for regulation under this paradigm was the achievement and maintenance of a margin of safety from a human health standpoint for exposure to environmental pollutants.

From a human health standpoint, the margin of safety factor is typically one-half, one-third, or one-tenth of the concentration at which best evidence indicates a critical health effect occurs. Moreover, such standards are often set to protect sensitive members of a population, for example, those with respiratory illnesses. An extreme form of the margin of safety approach is the so-called Delaney Clause, which sets a threshold of zero for any agent known to cause cancer at any level of exposure.[10]

The assumptions underlying this approach to analysis were first and foremost that environmental pollutants had significant impact on human health and on the natural environment. A further assumption was that causes and effects in the environment could be reasonably well tied together. The objective of policy was to reduce pollution to a point where a margin of safety for health and natural environmental impact could be maintained.

Two things happened that began to undermine the margin of safety approach. First, many of the most extreme forms of environmental pollution were brought under control and eliminated under this analytical regime. Visible emissions were recognized, identified, and controlled.

A second development was that detection technology became much more powerful, so that very small concentrations of pollutants could be identified in the field. Because of these more refined and sensitive detection techniques, the objective of reducing pollutant levels below detectable thresholds became an increasingly costly enterprise. Thus, better detection put the

---

[10] The Delaney Clause (Food Additives Amendment of 1958, P.L. 85-929, 72 *Stat.* 1784) directed the Food and Drug Administration to ban any food additive that causes cancer in humans or animals, regardless of the statistical risk.

88 FLEXIBLE URBAN TRANSPORTATION

margin-of-safety paradigm on a collision course with larger economic forces, since eliminating a pollutant below measurable thresholds became essentially cost prohibitive or impossible.

As with indicative planning in general, a concern about margin of safety analysis is that the marginal cost of achieving those indicated levels is not taken into consideration. Neither is the marginal benefit of one threshold well defined relative to slightly higher or slightly lower levels of the threshold. Thus, such margin of safety analysis ignores much of the marginal character inherent in systems with distributed and varied activity patterns like transportation.

Benefit–Cost Analysis

Benefit–cost analysis addresses some of the concerns about indicative planning insofar as it takes account of both the benefits and the costs of a particular course of action, which in turn reflects local conditions. Its foundation is that decisions should be based on the analysis of explicit alternative courses of action, guided by the relative value of each alternative's benefits and costs. Only actions with benefits that exceed costs should be undertaken, and in the presence of a budget constraint, project selection should maximize net benefits. Because benefit–cost analysis is a well established, only a cursory introduction to its basic techniques and recognized strengths and weaknesses is presented here.[11]

Because the temporal distribution of benefits and costs may vary, the conventional approach is to select a discount rate and to discount benefits and costs to "present value" on the basis of that discount rate. For a project with a planning horizon of n years, using a discount rate of i, the "net present value" (NPV) is defined as

$$NPV = \sum_{t=1}^{n} \frac{benefit_t - cost_t}{(1+i)^t}$$

One of the great strengths of the benefit–cost approach is that it captures some of the marginal dimensions of the tradeoffs between benefits and costs, an improvement over the indicative approach. Another strength is that it provides a framework for assessing impacts at a societal level, including on both those who are parties to a transaction or decision, and those who are affected by that decision but are not parties to it.

Benefit–cost analysis is subject to a number of well-known limitations, however. It fails to adequately treat distributional equity, uncertainty, irreversibility, and nonpecuniary impacts.

---

[11] For a concise treatment, see David Leo Weimer and Aidan R. Vining, *Policy Analysis: Concepts and Practice*, 3rd ed. (Upper Saddle River, N.J.: Prentice Hall, 1998), chap. 12. A more thorough treatment is Edward M. Gramlich, *Benefit–Cost Analysis of Government Programs* (Englewood Cliffs, N.J.: Prentice Hall, 1981).

Distributional Equity

The distribution of costs and benefits—that is, who benefits and who pays—introduces significant complications to benefit–cost analysis. While a decision according to benefit–cost criteria theoretically maximizes social welfare, it largely ignores how an alternative affects the distribution of those benefits across various subgroups of the population. Projects, especially large projects, often have impacts that are widely distributed over income groups, ethnic groups, and other subpopulations, over space (including crossing national and jurisdictional boundaries), and over time spans that may cross generations. The appropriate accounting of such distributions and their systematic incorporation into decision making raises many difficulties. Moreover, costs must usually be realized before benefits can be generated. In recent decades, public policy has increasingly emphasized the *ex ante* assessment of costs in the form of environmental impact statements and, some suggest, possible neglect of the assessment of benefits.[12]

Theoretically, it is possible for the beneficiaries of a project to compensate those who are made worse off by it for their losses and still come out ahead, since by definition benefits exceed costs. But in practice the benefits and costs of a project often fall on different groups and no compensation occurs, due to transaction costs or to project externalities. For example, how might suburban drivers compensate those living, working, or passing time near the highway? Who has standing for such benefits? The issues are thorny.

Intergenerational equity may be an issue as well, when benefits of a course of action fall on one generation and the costs on a different generation. Focusing analysis on the "present" value of investments in some ways gives a preference to the present generation vis-à-vis future generations. A justification for such a preference is the assumption that future generations are likely to be better off due to increases in productivity and knowledge and can take care of themselves. The present generation does most for them by creating as much wealth as possible now in order to bequeath it to them.

An emphasis on stewardship would characterize environmental resources as scarce, and part of the "wealth" passed on to future generations as an undisturbed environment. From this standpoint, projects or alternatives should be evaluated not on the basis of their present value but instead on their value to the present and all future generations. One could get at this, for example, by assembling a hypothetical panel of representatives from all generations where representatives are under a "veil of ignorance" about which generation they belong to.[13]

---

[12] Richard R. Mudge and Cynthia S. Griffin, "Approaches to the Economic Evaluation of IVHS Technology," in *Transportation, Information Technology and Public Policy*, Proceedings, IVHS Policy: A Workshop on Institutional and Environmental Issues, ed. Jonathan L. Gifford, Thomas A. Horan, and Daniel Sperling (Fairfax, Va.: Institute of Public Policy, George Mason University, 1992), 97-126.

[13] John Rawls, *A Theory of Justice* (Cambridge, Mass.: Harvard Belknap, 1971), Erhun Kula, "Discount Factors for Public Sector Investment Projects Using the Sum of Discounted Consumption Flows: Estimate for the United Kingdom," *Environment and Planning A* 16 (1984): 689-94; Bruce E. Tonn, "500-Year Planning: A Speculative Provocation," *Journal of the American Planning Association* 52, no. 2 (Spring 1986): 185-93.

The appropriate discount rate for assessing tradeoffs over time is not obvious and has been the subject of extensive discussion. Until recently, interest rates observed in the capital markets typically consisted of market expectations about several elements: time value of money, inflation, tax rates, and currency valuations. The recent introduction of inflation-protected bonds in Canada (called Canadian Real Return Bonds) and the U.S. (Treasury Inflation Protection Securities, TIPS) has provided a much better, albeit still imperfect, market indicator of rate of return. TIPS are ten-year, Treasury-backed securities. Each year, the holder of the bond receives interest earned on the bond plus (or minus) any inflation (or deflation) adjustment to the face value of the bond, based on the consumer price index (CPI) for all urban consumers. The U.S. successfully auctioned $7 billion in notes at a yield of 3.449 percent in January 1997, significantly higher than had been expected.[14]

Standard benefit–cost analysis utilizes the concept of externalities, that is, costs and benefits that fall upon those who are not party to a particular transaction. Congestion externalities are common in transportation projects (and other so-called club good problems) and pose particularly difficult problems. Typically one user imposes congestion on other users for which he does not compensate them; that is, he creates an externality. In the highway setting, this congestion externality manifests itself in the impacts of merging into a traffic stream upon entry to a highway, the occupation of scarce road space during travel, the turbulence caused by any lane changing or weaving during travel or preparation for exiting, and disruption of through traffic if exit ramps back up onto the main line. Because so much of the congestion externality derives from the impact of entering and exiting the traffic stream, short trips tend to have a disproportionately high externality impact per mile.

A frequently suggested (but rarely implemented) remedy for dealing with congestion externalities is some form of congestion pricing, whereby a pricing scheme is put into place that internalizes the congestion externality. Prices are set such that traffic is diminished sufficiently to prevent congestion, maintain free-flowing traffic, and increase social welfare.

While conceptually appealing, congestion pricing leaves a number of important questions unanswered. What level of congestion or free flow on the facility is socially optimal, and how should that be determined? Should it be done through the imposition of technical (and perhaps subjectively motivated) criteria like maintaining free flow (a performance criterion)? Keeping traffic below a given vehicle/capacity ratio (a technical criterion)? How should prices be set to incorporate the disproportionate impact of short trips? Should prices be set uniformly per mile, per trip, or some combination? How should travel time be incorporated into the pricing scheme? In the absence of a set of competing facilities with different price-congestion combinations for consumers to choose among, the socially optimal level of congestion is not self-evident. Any particular level of congestion will reflect a large dose of subjective judgment.

---

[14] TIPS rates still reflect expectations about taxes and currency valuations. Also, any bias in the CPI would also affect TIPS rates. The ten-year term of the bonds is also shorter than the duration of most transportation infrastructure investments. See Gregory Zuckerman, "Inflation-Linked Bond's Debut Bolsters Treasury Department," *Wall Street Journal*, January 30 1997; J. Ziegelbauer, "Taxes Lurk in New Inflation-Adjusted Treasury Bonds," available at www.gt.com/gtonline/tax/tba/dec96a.html, *Tax and Business Adviser* (Grant Thornton) 9, no. 4 (November/December 1996).

An ongoing project on State Route 91 in Orange County, California, promises to shed some light on this issue. For 10 miles along State Route 91 a toll facility has been built in the median of an existing freeway that offers reduced fees to vehicles carrying multiple occupants. Tolls vary depending on conditions on the parallel freeway. When the freeway is heavily congested, the express lane price is higher, and vice versa. A study has monitored the road conditions and traveler options before and after the implementation of this program. Preliminary results indicate that many travelers discriminate in judging when the time savings justify paying the toll (about 45 percent of the peak-period travelers were reported to use the facility once a week or less).[15]

Moreover, one school of economic thought questions the validity of the whole concept of externality. The problem, according to this line of reasoning, is that asserting the existence of an externality can only follow from an assumption of a pure, hypothetical world where all parties affected—however minimally—by an action are rightly compensated for it. In reality, only the most significantly affected stakeholders to a decision usually have enough at stake to participate in a transaction. Thus, concern about externalities should instead be directed at reducing transaction costs, or correcting outcomes that derive from excessive transaction costs.[16]

Sources of Uncertainty

A second major limitation of benefit–cost analysis derives from the often considerable uncertainty about both costs and benefits. There is an astonishing variety of classification schemes for uncertainty. Morgan and Henrion's typology identifies statistical variation, subjective judgment, linguistic imprecision, variability, inherent randomness, disagreement, approximation, and model formulation. Methods for addressing uncertainty in decision making are addressed later in the chapter.[17]

Statistical variation is the uncertainty that arises from the random error in direct measurements of a quantity, and it is the most commonly studied form of uncertainty. No measurement is exact; hence there is an irreducible random error in each measurement of a quantity. Successive repetitions of a measurement can allow that measurement error to be reduced to an arbitrary minimum.

Subjective judgment may also be the source of significant uncertainty in benefit–cost analysis. It is especially problematic when considering the stability of trends over time, both *ex ante* and

---

[15] ITE Taskforce on High-Occupancy-Tolls (HOT-Lanes), "High Occupancy Tolls (HOT-Lanes) and Value Pricing: A Preliminary Assessment," *ITE Journal* (June 1998): 30.
[16] Ronald Coase, "The Problem of Social Cost," *Journal of Law and Economics* 3, no. 1 (1960): 1-44; James M. Buchanan and William Craig Stubblebine, "Externality," *Economica* 29 (November 1962): 371-84; Carl J. Dahlman, "The Problem of Externality," *Journal of Law and Economics* 22 (April 1979): 141-62, reprinted in *The Theory of Market Failure: A Critical Examination*, ed. Tyler Cowan (Fairfax, Va.: George Mason University Press, 1989), 209-234.
[17] M. Granger Morgan and Max Henrion, *Uncertainty: A Guide to Dealing with Uncertainty in Quantitative Risk and Policy Analysis* (New York: Cambridge University Press, 1990), chap. 4.

ex post. The chief difficulty in *ex post* evaluation arises from judgments about what would have happened if a project or system had not been built, that is, the "counterfactual" problem. The problem is compounded in *ex ante* analysis when judgments determine which trends are to be held constant over time and which are to vary.

Another source of uncertainty is *linguistic imprecision,* which results from referring to events or quantities in imprecise language. The proposition that "Route 1 is congested" is imprecise because it does not specify what is meant by the term congestion, the time of day, or the specific location on Route 1. While fuzzy logic has provided some tools for formally addressing linguistic imprecision, the application of clear thinking to remedy linguistic imprecision is the approach preferred here.

The *variability* of quantities over space or time is another source of uncertainty. In some cases the frequency distribution for a particular quantity may be well defined, for example, the number of vehicles passing a particular point on a specific route during a specified period of time. In such cases it is legitimate to represent uncertainty about this quantity with a probability distribution with the same characteristics as the parent frequency distribution. It is also legitimate to combine, say, categories of roads into groups with similar frequency distributions, for example, principal arterials and collectors.

Variability is different from uncertainty about the nature of the frequency distribution that applies to a particular quantity, however, and it is important to recognize the difference. For example, in considering the impact of a new traffic control device on drivers, there may be uncertainty deriving from variability among drivers of different ages, as well as scientific uncertainty about the impact of the device on a specific age group. Disaggregating and computing impacts on each age can reduce the uncertainty due to variability. The scientific uncertainty can only be reduced by further research.

Inherent randomness, or aleatory uncertainty, from the Latin *alea* for a die or dice, is uncertainty arising from the inherent character of a process or quantity. Its distinguishing characteristic is that it is not reducible, even in principle. Heisenberg indeterminacy, from the dominant paradigm of quantum physics, is an example. It is impossible even in principle to measure exactly both the position and velocity of a particle.

Disagreement about the nature of a quantity or process can also be a source of uncertainty. While a consensus among experts might eventually emerge after years or decades of debate, in the moment of analysis of a particular project or course of action, disagreement about fundamental processes can be a considerable source of uncertainty. For example, clinical practitioners, inhalation toxicologists, and other specialties may differ subtly or sharply in their assessment of the health impacts of air pollution.

An analytical model is necessarily a simplified version of the complex reality it seeks to represent, which gives rise to *approximation uncertainty*. The physical resolution of a model— of an urban area consisting of so many traffic zones that generate or attract trips, for example—

and the temporal resolution—into hours, days, months, or years—represent necessary simplifications of the complexity of a real world space–time continuum.

Finally, *model form* introduces fundamental uncertainties to analysis. As Morgan and Henrion state, *"Every model is definitely false.* [However] we may be able to say that one model is *better* than another...." Model form uncertainty differs from approximation uncertainty in that the analyst may systematically assess the impact of the latter by, for example, testing the sensitivity of model estimates to changes in resolution. However difficult such testing may be, it is logically possible. Model uncertainty, on the other hand, is much less tractable.[18]

Two sources of uncertainty that are important in transportation do not fit neatly into Morgan and Henrion's typology. One is uncertainty that derives from incomplete knowledge of cause and effect within the domain of a particular problem. Demand for transportation, for example, is typically derived from the demand for goods or services that are available at a destination. Another source of demand is transportation for the sake of transportation such as hiking and bicycling, joy riding or cruising on Main Street on a Saturday night. As a result, the causes of demand for transportation facilities and services arise from a broad array of social and economic conditions, including the spatial distribution of goods and services, income, alternative opportunities for consumption and production, the quality and cost of communications, to name just a few. To have a complete understanding of the causes of transportation demand, it would be necessary to have a complete understanding of these broader social and economic processes, which is beyond the capacity of analytical models.

A second source of uncertainty important in transportation derives from variations in preferences both across individuals and over time. As individuals and organizations learn how to use transportation in different ways, their preferences and the value associated with the use of facilities and services changes in ways that may be difficult to predict. Even if the physical impacts of a project were known with certainty, for example, their economic value might be highly uncertain and subject to change over time. As incomes rise, for example, demand for leisure activities tends to rise, increasing their price relative to other activities. If a project, say a highway, will destroy a park but provide near-term economic benefits that exceed that park's current economic value, rising productivity and incomes may bring a time when the economic value of that park as a recreational site may exceed the value of the benefits accruing from the highway. If the park consists of natural features created in geologic time frames that are not reproducible by man, then it would be preferable from a present value standpoint to keep the park and forego the benefits of the highway.[19]

A combination of statistical variation and subjective judgment, which might be called *information quality,* is another source of uncertainty. Information quality takes into account the extent, precision, and timeliness (or vintage) of information available for making a decision. In the analysis of many transportation investment choices, the quality of data may be poor. For

---

[18] Morgan and Henrion, *Uncertainty*, 68, emphasis in the original.
[19] John V. Krutilla and Anthony C. Fisher, *The Economics of Natural Environments: Studies in the Valuation of Commodity and Amenity Resources*, rev. ed. (Washington, D.C.: Resources for the Future, 1985).

many projects, budgets for new data collection and travel surveys are limited. It is possible to use older data sets, but doing so embraces an underlying assumption that the vintage of the data does not significantly affect its accuracy—often a heroic assumption.

Intelligent transportation system (ITS) technologies have introduced better sensors, which can also improve the information available about the "real world" for planning and analysis applications. To date, most attention has focused on operational applications of the data generated by these sensors in, for example, incident detection and management, fleet communication and dispatching, and automated vehicle control. Of course such technologies can also improve the quality of data available for strategic applications such as planning and analysis. The speed with which the new data is incorporated into planning and decision making, however, often tends to be slower than in operational applications. As a result, the extent to which ITS will improve information quality for planning remains to be seen.

Two examples that illustrate the extent of uncertainties about transportation's impacts are the significant U.S. investments in railroads during the nineteenth century and the Interstate highway system in the twentieth. Economic history had a showdown in the 1960s over the magnitude of the impact of nineteenth-century railroads. Robert Fogel, who later won the Nobel prize for economics, concluded that railroads contributed only about 0.6 percent per year to growth in gross domestic product during the nineteenth century. He argued that, with technological innovation and increased investment, canal transport would have become much more productive if railroad technology had not been available. That contribution, compounded over a century, would have increased overall output by 82 percent. But other analyses using different methodologies found that the contribution had been much greater—4 percent per year—which compounded over a century would have increased economic output over fifty-fold.[20]

Another example of the challenge of counterfactual analysis is Friedlaender's assessment of the U.S. Interstate highway system, which concluded that its benefits outweighed its costs. But she acknowledged that she could not systematically account for the ripple effects of the reorganization of the economy at large that resulted from it, such as the shift from rail to truck in production and distribution processes.[21]

The difficulties in assessing impacts *ex ante* are even greater. *Ex post* analysis requires subjective judgment about one state of nature (that is, the road not taken) and measurement error about current and historical conditions (the course of action that was chosen). *Ex ante* analysis faces subjective uncertainty about all states of nature under consideration (that is, about all alternatives, including doing nothing), as well as measurement error about existing and past conditions.

---

[20] Robert W. Fogel, *Railroads and American Economic Growth: Essays in Econometric History* (Baltimore: Johns Hopkins University Press, 1964); Albert Fishlow, *American Railroads and the Transformation of the Antebellum Economy*, Harvard Economic Studies, 127 (Cambridge, Mass.: Harvard University Press, 1966); David Hounshell, personal communication (January 29, 1997).
[21] Ann F. Friedlaender, *The Interstate Highway System: A Study in Public Investment* (Amsterdam: North Holland, 1965).

While *ex ante* analysis can certainly use forecasts of future conditions, the uncertainty introduced by forecasts increases with the magnitude of the change under consideration. If changes are marginal, it may be relatively easy to array the range of possible responses using probabilistic techniques such as Monte Carlo simulation. When changes are significant, however, forecasts become much less useful, since current and past behavior are much less reliable guides to future system behavior.

The dilemma is that the less a project is likely to have major effects, the easier it is to assess its impact. Any project that is likely to induce changes in the organization of social and economic activity becomes much more difficult to analyze through benefit–cost analysis.

Some of this broad array of uncertainty can be addressed through the use of more sophisticated analytical techniques like probabilistic risk assessment, discussed further below. Other aspects, however, arise because of the complex and adaptive behavior of the environments into which the projects are incorporated. Urban transportation infrastructure investments often produce benefits that are extremely hard to predict due to the complex nature of the environments in which they operate. It is extremely difficult to predict exactly what will happen with or without a particular facility, and thus difficult to assess the benefits and costs of that project. This more fundamental unpredictability poses a particularly serious dilemma for the systematic evaluation of large-scale infrastructure projects such as major new facilities, large-scale crossings, and wholly new systems. All of these types of uncertainty limit the value of estimates of benefits and costs, which are the foundation elements of benefit–cost analysis.

Irreversibility, Non-Pecuniary Impacts, and Other Limitations

A third concern about benefit–cost analysis is the significant element of irreversibility in much infrastructure investment; costs, once realized, cannot be recaptured. A decision to defer a project, however, is in many respects reversible. The benefits that the project would have created in the interim would be lost, although presumably the funds that would have been invested in it would be earning benefits elsewhere. Also, conditions might improve in the interim, technology might get better or cheaper, demand might shift (i.e., different preferences or alternatives), and so forth. Recent analyses of benefit–cost analysis that take into consideration the irreversibility of an investment suggest that benefit–cost ratios on the order of three or four (i.e., expected benefits exceeding expected costs by a factor of three or four) are more appropriate for projects with large irreversible investments. This subject is pursued in greater detail in the section on investing under uncertainty.

A fourth problem with the benefit–cost approach is that costs and benefits are not all conveniently denominated in monetary units, the multi-attribute decision-making problem. Air pollution, global warming, neighborhood disruption, scenic and landscape effects, safety effects—all of these pose difficulties when a project requires tradeoffs along more than one dimension, which all projects do. This is not an insuperable problem by any means. Multi-attribute methods abound. Some use various methods for identifying weights for the various criteria in order to collapse them into a single measure. Other approaches identify thresholds or efficiency frontiers that help sort out the most efficient set of alternatives, such as data envelopment analysis (DEA).

A fifth limitation is that benefit–cost analysis is sensitive to the selection of alternatives for analysis. The quality of the decision is fundamentally dependent on the proper choice of alternatives. The best of a poor set of alternatives is a poor choice. Yet an objective assessment of the overall quality of a set of alternatives is difficult to achieve.

And finally, the cost of the analysis itself can be very high. Because of bounded rationality, the net benefit of further analysis is not systematically knowable. The completely thorough analysis has infinite cost.

As a result, the conditions under which benefit–cost analysis is most useful are those where project impacts are denominated in dollars and are fairly certain, where a discount rate can be determined unambiguously, where impacts are largely reversible, where equity impacts are negligible, and where structural economic impacts are unlikely.

Probabilistic Risk Assessment

A useful extension of the benefit–cost model attaches to design values probability distributions instead of point estimates and propagates the resulting uncertainty through the decision model to derive probability distributions for selected performance estimates (e.g., net benefits).[22] Such probabilistic models have been implemented for a number of problems, including some applications to transportation and infrastructure.[23]

The impact of tailpipe emissions on human health, for example, involves several important probabilistic events. First of all, there is enormous variability in the vehicle fleet itself. Some vehicles emit large quantities of pollutants, others small. These variations depend, among other things, on the extent of "cold starts," vehicle speeds, and the frequency of acceleration and braking. The actual fate of a pollutant once it is emitted is also highly variable. Some will disperse into the atmosphere and ultimately degrade; others will bioconcentrate or may have short-term consequences.[24]

Once in the environment, the next concern from a human health standpoint is the levels of exposure that humans face as a result of that emission. Again, depending on the fate of the particular pollutant, there may be an enormous amount of variability in the actual level of exposure of particular individuals. Finally, individuals vary in their response to exposure to a

---

[22] Kingsley E. Haynes et al., "Planning for Capacity Expansion: Stochastic Process and Game-Theoretic Approaches," *Socio-Economic Planning Science* 18, no. 3 (1984): 195-205.
[23] The firm Hickling, Lewis and Brod has developed a proprietary package, the Risk Analytic Package (RAP), and used it in a number of applications, including airport expansion and airport bond risk assessment.
[24] Bioconcentration occurs when a toxin accumulates permanently in a bodily tissue, rather than being destroyed in digestion or excreted. DDT, for example, bioconcentrates in fat.

given pollutant according to a "dose response" relationship that may vary from individual to individual. That is, the health impact of a particular dose of a particular pollutant will often have a high level of variability that depends on the underlying health of the individual concerned, his or her genetic makeup or genetic predisposition to particular conditions, as well as a number of other factors.

Under an indicative planning approach determining a margin of safety, the analyst would select a typical or criteria individual or condition for each of the probabilistic events. For example, an analyst might pick a typical automobile emission profile, a typical environmental fate characterization, a typical level of exposure to the pollutant, and a person with a typical dose response function and thereby assess the health impact on a typical basis. "Typical" might mean the median individual or condition. "Criterion" suggests a more conservative assumption, for example, the 90th percentile of the dose–response distribution.

Probabilistic risk assessment takes a different approach. For each of those probabilistic events or conditions, it specifies probability distributions instead of single typical or criteria numbers. Hence, it would require as input a probability distribution to characterize the emissions profile of the vehicle, a probability distribution to characterize the fate of each pollutant, a probability distribution to characterize the dose–response relationship in a range of individuals. Factoring together all of those probability distributions using Monte Carlo or other simulation techniques would yield a probability distribution for the human health impact of tailpipe emissions.

Perhaps the most straightforward applications of probabilistic risk assessment utilize "add-in" packages to standard spreadsheet packages, such as @RISK from Palisade Software.[25] This add-in allows one to specify for any cell in a spreadsheet model a particular probability distribution with user-specified parameters. One might, for example, specify a cell as normally distributed with a mean of 10 and a standard deviation of 7, or as having a Poisson distribution with a mean of 3.

The add-in then assigns random values to the specified cells and calculates the model. Successive executions of the model, with successive random assignments to the specified cells, allow the program to generate a probability distribution for any outcome variable of interest.

The strength of the probabilistic risk assessment approach is that it represents the probabilistic nature of the problem domain much more accurately than a deterministic model. It can provide a much better understanding of the nature of the interactions involved. Given the pervasiveness of uncertainty and probabilistic character in the problem domain, this is a significant strength.

Its weaknesses, on the other hand, are several. First of all, it is extremely information intensive. It requires good information about the nature and shape of a wide variety of probability distributions involved in a particular problem. Frequently, good information about the nature of these distributions is not available, making it necessary to assume the nature and shape of the

---

[25] The software package Analytica, by Lumina Decision Systems (www.lumina.com), offers a more formalized and systematic environment for such modeling.

probability distribution, which in turn can introduce a degree of subjectivity into the analytic process.

A second problem is that although the probablistic risk assessment approach is sophisticated in its ability to represent complex problems, non-technical audiences often find it difficult to understand probability distributions and probabilistic phenomena. This technical inscrutability is particularly unfortunate because elected decision makers who have the highest level of accountability and legitimacy to the electorate often are not trained in the statistical tools and techniques necessary to understand probabilistic risk assessment. The technical experts who are well trained in the statistical content of such models are also the ones who have to make value judgments that may affect the outcome of models.

A final difficulty with probabilistic risk assessment is that it cannot thoroughly represent "model uncertainty," that is, uncertainty about the actual structure of the underlying process being analyzed. For example, in analyzing the carcinogenicity of a particular chemical, laboratory tests often use high concentrations of a chemical to assess whether cancers are induced in a test subject, usually an animal. Without understanding the detailed process through which the chemical induces the cancer, however, it is difficult to extrapolate the probability of inducing a cancer due to exposure at a much lower concentration of that chemical such as might be found in the everyday environment. Uncertainty about whether such extrapolation is appropriate cannot be adequately represented in the model.[26]

Life-Cycle Cost Analysis (LCCA)

Another extension of benefit–cost analysis is life-cycle cost analysis (LCCA). The motivation for life-cycle cost analysis is that decision makers often choose project and acquisition alternatives with the lowest initial cost. The proper choice should be that alternative (or set of alternatives) that maximizes net benefits. The preference for least initial cost over maximum net benefit as a selection criterion persists for a number of reasons. First, many public agencies face chronically low budgets, far below their own assessments of their requirements for either capital investment or operations and maintenance. In the face of such shortages, agencies may simply opt for the lowest initial cost as a mechanism for maximizing their ability to address chronic shortages.

From a technical standpoint, of course, it is possible to incorporate budget constraints into an analysis, in which case maximum net benefits would remain the proper decision criterion. But it may not be easy to get decision makers to be explicit about their assessments of budget shortfalls.

A second reason for the appeal of lowest initial cost grows out of the nature of public procurement processes in the U.S. Most procurement procedures allow losing bidders to challenge (or "protest") a contract award if they feel its selection was not based on fair and

---

[26] Morgan and Henrion, *Uncertainty*, esp. chap. 4.

objective criteria. Life-cycle cost estimates generally require forecasts of future operating costs as well as rates of technological innovation and obsolescence. Such forecasts can be somewhat subjective, and they may provide grounds for a successful contract challenge. Initial cost, on the other hand, is typically a more certain quantity that is specified in bid documents and therefore less open to challenge.

Again, uncertainty about future costs and opportunities can be incorporated systematically into benefit–cost analysis through the use of expected values or probabilistic risk assessment. But contracting officers often see a successful procurement as one that can successfully withstand challenge rather than one that best serves the program making the procurement. And the ability for such technically sophisticated techniques to stand up to a legal challenge is a risk that they may not be eager to take.

New contracting mechanisms such as "build-operate-transfer" (BOT) and "design-build-operate-transfer" (DBOT) have the effect of internalizing tradeoffs between design, construction, operating, and maintenance costs. As a result, they have the effect of moving towards life-cycle cost. However, such contracting techniques are not yet in widespread use in government provision of transport facilities and services.

Congress mandated the use of life-cycle cost analysis for bridges, tunnels, and pavements in the Intermodal Surface Transportation Efficiency Act (ISTEA) in 1991, and for projects on the National Highway System (NHS) costing more than $25 million in 1995. Concern among states, Congress, and others about mandating LCCA for all projects led to its being recommended but not required, except for national highway system projects over $25 million. To foster adoption and implementation, the Federal Highway Administration has also developed decision support tools and training for the application of life-cycle cost analysis in pavement design, which include Monte Carlo simulation techniques.[27]

Least Cost Planning

Least cost planning (LCP), also called integrated resource planning (IRP), is a broader, more comprehensive form of decision making than traditional benefit–cost analysis. One key difference is the range of alternatives that are examined in a particular decision situation. While any alternative is admissible under benefit–cost analysis, traditional approaches often emphasize supply-side improvements, that is, alternatives focused on accommodating changes in demand through adjustments in supply. LCP, on the other hand, explicitly embraces both supply and demand alternatives. It also provides a framework for examining multiple modes. A second key difference is procedural. Benefit–cost analysis is heavily grounded in economic

---

[27] "Intermodal Surface Transportation Efficiency Act of 1991," P.L. 102-240, 1105 *Stat.* §§1024, 1025 (18 December 1991); "National Highway System Designation Act of 1995," P.L. 104-59, 109 *Stat.* §303 (28 November 1995). U.S. Federal Highway Administration, "Life-Cycle Cost Analysis," final policy statement, *Federal Register* 61, no. 182 (September 18, 1996): 49187-91; James Walls III and Michael R. Smith, *Life-Cycle Cost Analysis in Pavement Design: Interim Technical Bulletin*, U.S. Federal Highway Administration no. FHWA-SA-98-079 (Washington, D.C., September 1998).

theory and often dominated by technical considerations and technical experts. LCP, on the other hand, seeks to be broadly participatory and embrace a more comprehensive range of perspectives.[28]

The origins of LCP lie in the electric utility industry, where many state regulatory commissions adopted or mandated LCP in order to foster consideration of multiple approaches to energy supply and demand. LCP, for example, could incorporate energy conservation programs as an alternative to new power generation and transmission facilities, as well as co-generation, solar, wind, and other supply approaches. In electric power, the term "least cost planning" has now given way to "integrated resource planning" or "strategic resource planning."[29]

Investing under Uncertainty

The orthodox theory of investment that governs most infrastructure investment analysis and environmental benefit–cost analysis stipulates that if the net present value (NPV) of the decision is positive—that is, if the discounted stream of expenditures and revenues, or societal costs and benefits, is greater than zero—proceed with the decision.[30]

As discussed above, a serious limitation of the orthodox theory is that it deals crudely or not at all with three important characteristics of many decisions. First, most decisions or investments entail some degree of *uncertainties and irreversible impacts*. That is, once undertaken, costs are at least partially sunk and cannot be recovered. Second, there is *uncertainty* about the future costs and/or benefits of the investment. And, third, decision makers usually have some control over the *timing* of a decision, that is, they have some ability to defer an action until better information is available in the future.

These three characteristics are pervasive in infrastructure investment and environmental decision making. Infrastructure investments are highly irreversible. The salvage value of an infrastructure asset is usually only a small fraction of its construction cost. Many public infrastructure investments are virtually irreversible because of public reliance on the services they afford. Similarly, the decommissioning of a locally unpopular facility may be difficult or impossible to reverse if subsequent conditions merit its reconstruction. Environmental degradation may also be largely irreversible, at least when measured in human generations. The loss of a species may never be reversed. The loss of an old growth forest may take hundreds of years to regenerate.

---

[28] Patrick DeCorla-Souza, Jerry Everett, and Brian Gardner, "Applying a Least Total Cost Approach to Evaluate ITS Alternatives," paper presented at the Transportation Research Board 74th Annual Meeting, January 22-28, 1995, Washington, D.C.; Dick Nelson and Don Shakow, "Least-Cost Planning: A Tool for Metropolitan Transportation Decision Making," *Transportation Research Record* 1499 (1995): 19-27.

[29] David Berry, "The Structure of Electric Utility Least Cost Planning," *Journal of Economic Issues* 26, no. 3 (September 1992): 769-89; Stephen Connors, personal communication (1997).

[30] For an excellent introduction to the literature on investment under uncertainty and the problems it raises with traditional net present value analysis, see Avinash K. Dixit and Robert S. Pindyck, *Investment under Uncertainty* (Princeton, N.J.: Princeton University Press, 1994), from which this section draws heavily; for public policy applications, see Gilbert E. Metcalf and Donald Rosenthal, "The 'New' View of Investment Decisions and Public Policy Analysis: An Application to Green Lights and Cold Refrigerators," *Journal of Policy Analysis and Management* 14, no. 4 (Fall 1995): 517-31.

Uncertainty is also pervasive in infrastructure investments and environmental decision making: in the nature and extent of future demand for infrastructure services, in the conditions of a site that may affect construction costs, in the extent to which perturbations in ecosystems represent irreversible secular changes of state, and so forth.

The timing of infrastructure investments and environmental decisions is also important. Infrastructure investments can be deferred or pursued incrementally to reduce uncertainty about, say, demand for a new facility. Environmental decisions can be accelerated or deferred in light of uncertainty about the nature, extent, or irreversibility of environmental impacts.

A new theory of investing takes account of irreversibility, uncertainty, and timing by recognizing that the option of waiting to invest or decide is an opportunity cost that may be significant. The new theory values this option explicitly and incorporates it into the decision to invest now or to wait for better information. This section works through some simple examples to illustrate how option value can depart from the orthodox theory, adapting examples from Dixit and Pindyck.[31]

Consider an agency that is trying to decide whether to build a highway segment. The investment is completely irreversible—the highway can only be used to carry traffic, and if no traffic materializes, the agency cannot recover its investment. For simplicity, assume that the highway can be built instantly for cost I, and that trips using the highway are worth $P = \$1$ each. Currently the expected traffic is 200 trips, but next year the traffic will change. With probability q, it will rise to 300, and with probability $(1 - q)$, it will fall to 100. The traffic will then remain at this level forever. While these are strong assumptions, the conclusions they imply are fairly robust and are developed more fully in Dixit and Pindyck, from which the example is adapted.[32]

The general formula for the orthodox NPV is:

$$\mathrm{NPV} = -I + \sum_{t=0}^{\infty} \frac{D_0 \times P}{(1+i)^t}.$$

Assume a risk-free interest rate, i, of 10%, with $I = \$1600$ and $q = 0.5$. Assume also that demand, D, is unrelated to what happens to the overall economy.

Given these values, should the agency build the highway today, or should it wait a year and see whether the traffic goes up or down? Investing now, noting that the expected value of traffic is 200), yields

$$\mathrm{NPV} = -1600 + \sum_{t=0}^{\infty} \frac{200}{(1.1)^t} = -1600 + 2200 = \$600.$$

---

[31] Dixit and Pindyck, *Investment under Uncertainty*, chap. 2.
[32] Dixit and Pindyck, *Investment under Uncertainty*, 27-29.

It appears that the NPV of this project is positive. The current value of the highway segment, V_0, is equal to $2200, which exceeds the $1600 cost of the highway. Hence, it would seem desirable to go ahead with the highway.

This conclusion is incorrect, however, because the calculations ignore the opportunity cost of investing today rather than waiting and retaining the option of not investing should traffic fall. To illustrate this, calculate the NPV of waiting one year and then investing only if traffic increases. In this case

$$\text{NPV} = (0.5)\left[\frac{-1600}{1.1} + \sum_{t=1}^{\infty}\frac{300}{(1.1)^t}\right] = \frac{850}{1.1} = \$773.$$

(Note that there is no expenditure and no revenue in year 0. In year 1, the $1600 is spent only if traffic rises to 300, which will happen with probability 0.5.) Waiting a year before deciding whether to invest in the highway yields a project NPV of $773, whereas it is only $600 for investing in the highway now. Clearly it is better to wait than to invest right away.

Note that if the choice were only between investing today or never, it would be better to invest today. If there were no option to wait a year, there would be no opportunity cost of killing that option. Likewise, if it were possible to liquidate the $1600 investment and fully recover the funds, it would make sense to invest today. Three things are necessary to introduce an opportunity cost into the calculation: irreversibility, uncertainty about future demand, and the option to wait.

The value of the option is simply the difference in the two NPVs: $773 - $600 = $173. That is, one would be willing to pay up to $173 more for an alternative that includes the option.

Varying Construction Cost

An alternative way to value this flexibility is to ask how much more one would be willing to pay in construction cost I to have an investment that is flexible rather than inflexible. To answer this question, find the value of I, call it $\bar{I}$, that makes the NPV of the project when waiting equal to the NPV when $I = \$1600$ and investing now. Substituting $\bar{I}$ for the $1600 and substituting $600 for the $773 in the previous equation:

$$\text{NPV} = (0.5)\left[\frac{-\bar{I}}{1.1} + \sum_{t=1}^{\infty}\frac{300}{(1.1)^t}\right] = \$600$$
$$\bar{I} = \$1980.$$

In other words, the opportunity to build a highway *now and only now* at a cost of $1600 has the same value as an opportunity to build the highway *now or next year* at a cost of $1980.

Two additional values of I are noteworthy. First, there is a value of I, I', at which it makes sense to invest today, that is, the option of waiting for a year sinks to zero. To find that value, equate the NPV of starting now with the NPV of waiting for one year and solve for I':

$$(0.5)\left[\frac{-I'}{1.1} + \sum_{t=1}^{\infty} \frac{300}{(1.1)^t}\right] = -I' + \sum_{t=0}^{\infty} \frac{200}{(1.1)^t}$$

$$I' = \$1284$$

Second, there is a value of I, I'', at which it makes no sense to invest even if demand rises to 300. To find that value, equate the NPV of waiting for one year to zero, and solve for I'':

$$(0.5)\left[\frac{-I'}{1.1} + \sum_{t=1}^{\infty} \frac{300}{(1.1)^t}\right] = 0$$

$$I'' = \$3300$$

Hence, depending on the value of I, we should either undertake the project immediately if $I < 1284$, wait for a year to see if demand rises to 300 if $1284 < I < 3300$, or abandon the project altogether if $I > 3300$, as illustrated in Figure 5-2.

Figure 5-2
Traditional NPV and Option Value

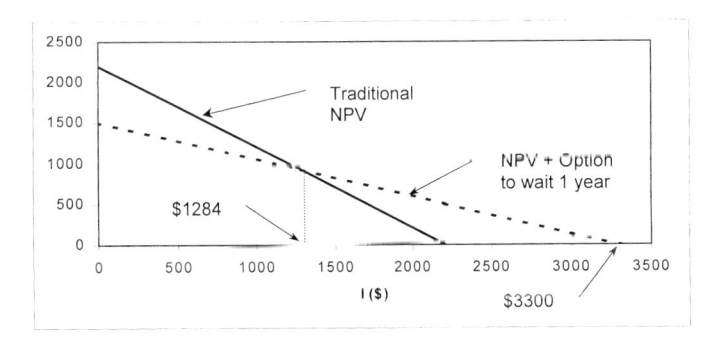

Varying Demand

One can also examine how the value of the project and the option to wait a year vary with the level of demand. In the original example, the expected value of demand, D_0, was 200, since there was a 50–50 chance that demand would rise to 300 or fall to 100 in year 2. At what value of D_0, call it D_0', would one be willing to invest now even if demand were to drop by 50 percent in year 2? Intuitively, this means that the benefit to be earned in year 0 by investing now is sufficient to offset the value of the option to wait and see what happens to demand in year 2. Therefore, equate the NPV of waiting with the NPV of investing now and solve for D_0':

$$(0.5)\left[\frac{-1600}{1.1} + \sum_{t=1}^{\infty}\frac{1.5D_0'}{(1.1)^t}\right] = -1600 + \sum_{t=0}^{\infty}\frac{D_0'}{(1.1)^t}$$

$$D_0' = 249.$$

Similarly, to determine what value of D_0, say D_0'', would justify bypassing the project completely, set the NPV of waiting for one year to zero, and solve for D_0'':

$$(0.5)\left[\frac{-1600}{1.1} + \sum_{t=1}^{\infty}\frac{1.5D_0''}{(1.1)^t}\right] = 0$$

$$D_0'' = 97$$

Hence, if $D_0 > 249$, invest immediately; if $97 < D_0 < 249$, wait for a year and invest if the price rises; and if $D_0 < 97$, abandon the project.

Increasing Uncertainty over Demand

One may also vary the extent to which demand might rise or fall. The original example assumed demand would rise or fall by 50 percent. How does the value of the option vary if that variance rises or falls? Assume the probability of a rise in demand, q, is 0.5, but that the demand will either rise or fall by 75 percent in period 1. Then

$$\text{NPV} = 0.5\left[\frac{-1600}{1.1} + \sum_{t=0}^{\infty}\frac{1.75 \times 200}{(1.1)^t}\right]$$

$$= \$1023.$$

The value of the option is $1023 − $600 = $423, compared with the $173 in the original example. Why does the value of the option value increase with the variance of future demand? Because it increases the upside potential payoff from the option, leaving the downside risk unchanged, since we will not exercise the option if demand falls.

"Bad News Principle"

An important implication for infrastructure investment of option value is the "Bad News Principle." By modifying the example by allowing both the probability of an upward move in demand, q, as well as the magnitude of the upward and downward moves to vary, it is possible to determine how "good news" (an upward move) and "bad news" (a downward move) separately affect the critical demand level, D'_0, that warrants immediate investment. (In the foregoing examples, the magnitude and probability of upward and downward moves increased and decreased together.) It turns out that the critical demand level, D'_0, depends *only* on the size of the downward move, not on the size of the upward move. The reason is that the ability to avoid the consequences of "bad news" makes waiting desirable.

Suppose that the initial demand level is D_0, but in period 1 the demand level becomes

$$D_1 = \begin{cases} (1+u)D_0 & \text{with probability } q \\ (1-d)D_0 & \text{with probability } (1-q) \end{cases}$$

Keeping the cost of the investment at I, the NPV of investing now is

$$\text{NPV} = -I + D_0 + q\sum_{t=1}^{\infty} \frac{(1+u)D_0}{(1.1)^t} + (1-q)\sum_{t=0}^{\infty} \frac{(1-d)D_0}{(1.1)^t}$$
$$= -I + 10[1.1 + q(u+d) - d]D_0$$

On the other hand, the NPV of waiting is

$$\text{NPV} = \frac{1}{1.1}\{q\max[0, -I + 11(1+u)D_0] + (1-q)\max[0, -I + 11(1-d)D_0]\}.$$

Clearly, the point of indifference between investing now and waiting is in the range of D_0 where investment in period 1 is warranted if the price goes up but not if it goes down. In this case, the NPV of waiting simplifies to

$$\text{NPV} = \frac{q}{1.1}[-I + 11(1+u)D_0].$$

Equating the NPV of investing now and the NPV of waiting and solving for D_0 gives

$$D'_0 = I\left(\frac{1}{1.1}\right)\left(\frac{0.1 + (1-q)}{0.1 + (1-q)(1-d)}\right).$$

This equation has one important detail: D'_0 does not depend in any way on u, the size of an upward move. It only depends on the size of a downward move, d, and the probability $(1-q)$ of a downward move. Also, the larger is d, the larger is the critical demand level, D'_0; it is the magnitude of the possible "bad news" that drives the incentive to wait.

Moreover, if current net benefit can be negative and a society is contemplating a costly disinvestment or abandonment of a project, the bad news principle turns into a good news

106   *FLEXIBLE URBAN TRANSPORTATION*

principle: the size and probability of an upturn are the driving forces behind the incentive to wait to disinvest.

Scale versus Flexibility

Economies of scale can be an important source of cost savings. By building a large facility instead of two or three smaller ones, society may be able to reduce its average costs and increase social well-being. This suggests that society should respond to growth in demand for services by bunching its investments, that is, investing in new capacity only infrequently, but adding large and efficient facilities each time.

What should society do when there is uncertainty about demand (as there usually is)? If society irreversibly invests in a large addition of capacity, and demand grows only slowly or even shrinks, it will find itself holding capital it does not need. Hence, when growth of demand is uncertain, there is a tradeoff between scale economies and the flexibility that is gained by investing more frequently in small increments to capacity as they are needed.

This problem is important in road building. It is generally much cheaper per unit of capacity to build a four-lane facility than it is to build a two-lane facility. At the same time, road building authorities face considerable uncertainty about how fast the demand for a road will grow. Hence, it is important to be able to value this flexibility.

To adapt the earlier example, assume that the highway segment in question can be built either in a four-lane configuration at a cost of $I_0 = \$1600$ (Option A), or in a two-lane configuration at a cost of $I_o = \$1000$ (Option B). If built in a two-lane configuration, it could be expanded at a cost of \$1000 to four lanes in year two. Demand is uncertain. It is presently at 200, but with a 50–50 chance it may grow to 300 in year 1 or remain at 200.

The NPV of the four-lane option is

$$\text{NPV}_A = -1600 + 200 + \sum_{t=1}^{\infty} \frac{250}{(1.1)^t} = \$1100$$

The value of the two-lane option is:

$$\text{NPV}_B = -1000 + 200 + (.5)\sum_{t=1}^{\infty} \frac{200}{(1.1)^t} + (.5)\left(\frac{-1000}{1.1} + \sum_{t=1}^{\infty} \frac{300}{(1.1)^t}\right)$$

$$= -800 + 1000 - \frac{500}{1.1} + 1500$$

$$= \$1245$$

Hence, uncertainty about demand growth makes the option to build incrementally preferable, even at a loss of economies of scale. The total construction cost of Option B is:

$$1000 + \frac{1000}{1.1} = \$1909,$$

compared with Option A's $1600, a difference of $309, or 16 percent, and the option value present in Option B is $145 (1245-1100).

Extensions to this logic include varying the probability of demand increasing or decreasing, extending the uncertainty beyond a single period, and incremental projects, where the first few steps of a project reveal information necessary for assessing its total value, as in the case of preliminary engineering studies. These are not explored in detail here but are available in Dixit and Pindyck.[33]

Environmental Applications

Just as capital investment decisions may involve irreversible actions in the face of uncertainty where timing is discretionary, so may environmental decisions. Indeed, some of the most profound irreversibilities and uncertainties associated with capital investments in infrastructure are their environmental impacts.

Investing under Uncertainty: Summary

The examples used here to illustrate the importance of option value are subject to many simplifying assumptions. Dixit and Pindyck's book relaxes these assumptions and asserts the following general insights from the theory of investing under uncertainty:[34]

1. The value of the option to wait can be significant and may justify waiting rather than proceeding immediately.
2. Option value increases with the value of the sunk cost, I.
3. Option value increases with the degree of uncertainty over future demand (i.e., the variance of future demand), the downside risk being the most important aspect.
4. Conditions that merit killing an option to wait and proceeding immediately with a project often imply orthodox NPVs considerably higher than unity.
5. "When a project consists of several steps [and] the uncertainty about it will be revealed only as the first few steps of the project are undertaken, ... then these first few steps have information value over and above their contribution to the conventionally calculated NPV. Thus it may be desirable to start the project even if orthodox NPV is somewhat negative."

---

[33] Dixit and Pindyck, *Investment under Uncertainty*, chap. 2.
[34] Dixit and Pindyck, *Investment under Uncertainty*, 11.

6. "Investment on a smaller scale, by increasing future flexibility, may have a value that offsets to some degree the advantage that a larger investment may enjoy due to economies of scale."

These general insights from the new investment theory have profound implications for infrastructure investment. Developing a decision-making process that is appropriate under the common conditions of irreversibility, uncertainty, and discretion over the timing of investment is one of the central challenges addressed later in this book.

Participatory Decision Making

Both indicative planning and benefit–cost analysis depend heavily on technical expertise. Participatory decision making takes a sharply different view of how decisions should be made by giving authority to the views of a full range of stakeholders in a decision. For many transportation projects, stakeholders could include neighborhood groups near a planned facility, environmental interest groups, local chambers of commerce, labor unions, and associations of motorists, bicyclists, and transit users, as well as others.

Participatory approaches are based on the notion that those affected by a decision should have at least some say in it, regardless of whether their interests are adequately reflected in technical models and analyses. Recognizing that most decision making requires a considerable degree of subjective judgment, participatory decision making directly engages the affected parties to elicit their preferences about impacts. Participatory approaches take a number of forms. Most include some sort of group interaction by affected or potentially affected parties (households, firms, and associations).

Public participation in decision making is a form of direct democracy, as distinct from representative democracy. Public participation has become increasingly important in transportation decision making in recent decades. The emergence of public participation in part reflects the difficulty of the technically or scientifically based approach. It may also be attributable in part to a distrust of technical experts that derives from the Vietnam and Watergate era, and what some see as an increasing distrust in government, government lobbyists, and the influence of special interests. It also reflects recognition of the legitimacy of critical and persuasive discourse, as discussed earlier, in addition to analytical discourse in public choices about transportation facilities and services.

An example of a participatory process would be a citizens review board or citizens advisory board for the design of a major highway project. Such a committee was utilized, for example, in the design and construction of Interstate 70 through Glenwood Canyon, Colorado, discussed in the last chapter.

In the reauthorization of the national highway and transit programs in 1991, Congress placed considerable weight on increasing public involvement in transportation decision making.[35] In 1993, the Federal Highway Administration and the Federal Transit Administration revised their planning requirements for federally supported projects accordingly,[36] and developed a general policy on public involvement in transportation planning and decision making in 1995.[37]

The policy and regulations require states and metropolitan areas "to establish their own continuing public involvement processes which actively seek involvement throughout transportation decisionmaking, from the earliest planning stages, including the identification of the purpose and need, through the development of the range of potential solutions, up to and including the decision to implement specific solutions.... *[T]he public must receive assurance that its input is valued and considered* in decisionmaking so that it feels that the time and energy expended in getting involved is meaningful and worthwhile."[38]

One limitation of this approach is that those involved in the decision-making process are usually self-selected, and may have a strong personal interest in particular impacts or a bias for particular solutions. As a result, they may not be representative of the broader population. Reliance on such a self-selected group can result in an overrepresentation of extreme views, rather than a broader representation of the full range of views. This can in turn lead to biased decision making. In this sense, market-based approaches are much more representative of mainstream public preferences, since individuals express their preferences through behavior, rather than through a public decision process. But, as discussed earlier, they are also constrained by social and economic endowments.

A second shortcoming of the public participation approach is that participants may lack specific technical knowledge needed to understand design and analytic issues. In some cases, including technical experts in a decision-making process along with stakeholders can remedy such a deficiency. But it is at least conceivable that stakeholders might simply reject information that is technically or scientifically valid.

One remedy that has been used to blend the participatory approach with a technically based analysis combines risk analysis and public participation. One firm has specialized in the application of such a combined approach to transportation decision making. Stakeholders to a decision are educated about the technical content of the problem and participate in the development of a probabilistic risk analysis that represents all relevant aspects of a problem. In a decision about the expansion of an airport, for example, members of a citizens advisory board

---

[35] "ISTEA," §§ 1024, 1025, 3012.
[36] 23 CFR parts 450.212 and 450.316 (b) (1), as revised in U.S. Department of Transportation, "Statewide Planning; Metropolitan Planning," final rule, *Federal Register* 58, no. 207 (October 28, 1993): 58067, 58073.
[37] U.S. Federal Highway Administration and Federal Transit Administration, "Interim Policy and Questions and Answers on Public Involvement in Transportation Decisionmaking," notice, *Federal Register* 60 (January 27, 1995): 5463-68.
[38] Emphasis in original. U.S. Federal Highway Administration and Federal Transit Administration, "Interim Policy and Questions and Answers on Public Involvement in Transportation Decisionmaking," 5465, 6467.

made up of directly affected parties and interested parties were trained and educated about the nature of airport operations, the importance of aircraft size, the impact of runway length, future scenarios for the aircraft industry, etc. in a way that allowed them to make informed decisions about the relative benefits and costs of an airport expansion in comparison with other options, including the building of an additional airport. While such approaches have promise, they can also be extremely time-consuming for the participants.

***

This chapter has examined a broad range of transportation planning methodologies, decision-making concepts, and approaches relevant to transportation drawn from other fields. It has been necessarily brief in its treatment of many topics, but it has emphasized the pervasiveness of uncertainty in the decision-making environment, and tools and approaches for systematically addressing uncertainty. The next chapter brings the focus to highway and transit planning in the U.S., with a discussion of its evolution over the last century.

6

THE EVOLUTION OF TRANSPORTATION PLANNING

Highway and transit planning in the United States has been heavily influenced by the dominant role of federal funding for facilities and services. Indeed, considering all the nation's lifeline infrastructure systems—electricity, telecommunications, water supply, wastewater treatment, transit, and highways—the federal influence is perhaps greatest in highway and transit planning.

The previous chapter described three broad categories of planning. Under indicative planning, decisions are based on broad indicators of adequacy or sufficiency. Under benefit–cost analysis, decisions are based on estimates of net benefits. Under participatory planning, decisions are based on the views of stakeholders and affected parties.

Highway and transit planning in the U.S. has been dominated by indicative planning, which, in recent years, has begun to be displaced by participatory decision making. This chapter provides a brief overview of the major U.S. surface transportation planning initiatives of the last century. It begins with a description of planning in the early years of the Federal-Aid Highway Program, between 1916 and World War II. Next is a discussion of metropolitan transportation planning, which was closely associated with the development of the Interstate program and has become tightly intertwined with air quality planning. It then turns to national-scale planning, which includes a discussion of renewed initiatives for indicative highway classification, highway needs studies, and broader "condition and performance" studies. It also discusses research of the last decade on the linkage between transportation investment and economic performance at the national level.

HIGHWAY PLANNING PRIOR TO WORLD WAR II

In the early years, beginning in the 1920s, highway planning consisted of designating systems of federal-aid highways within the states, and then using federal aid and state funds to improve those roads to acceptable standards. The major planning activity consisted of designating the highways that would comprise the federal-aid system. States nominated a system and then negotiated with the Bureau of Public Roads to arrive at an accepted system. The 1921 highway act limited the length of a state's federal-aid system to 7 percent of its improved roads, which was intended to force the state to make tough choices and designate only those highways that were most important.

The Bureau's interest lay in ensuring that the system was internally connected—that is, with all links connected with other links in the system—with minimal "stub ends" (links connected at only one end), connected to the major centers of population and industry, and linked to highways in adjacent states. Traffic studies served as an important planning tool during this period, and the level of traffic on a road was an important indicator of its suitability for inclusion in the federal-aid system. Project-specific estimates of the benefits of bringing a road into compliance with a standard were not required and were usually not undertaken. Program-level justifications typically focused on outputs, like miles of roads improved to standard, as opposed to any economic benefits from those improvements.

During the Depression, the need for an economic rationale for the highway program led to a focus on jobs. Also, concern about the robustness of the national economy gave rise to a spate of regional development and planning exercises under the auspices of the National Resources Planning Board.[1]

METROPOLITAN TRANSPORTATION PLANNING

Urban areas posed planning problems that simple traffic studies could not address. Urban street networks were much more complex than rural networks. It was much more difficult to predict the effect of adding or improving an arterial, since traffic on adjacent streets could be expected to divert to an improved route once it was available. As a result, traffic counts alone were not an adequate indicator of where improvements were needed. Moreover, data on urban traffic movements was poor. Most cities did not begin to collect data on traffic systematically until the 1940s.[2]

The earliest metropolitan planning efforts relied on surveys of travelers' origins and destinations. Lines plotted on a map connecting origins and destinations, called "desire lines," served as indicators of demand intensity, with the density of coverage providing an indicator of the level of travel demand.

---

[1] See Patrick D. Reagan, *Designing a New America: The Origins of New Deal Planning, 1890-1943* (Amherst: University of Massachusetts Press, 1999); Jim Tomlinson, review of *Designing a New America: The Origins of New Deal Planning, 1890-1943*, by Patrick D. Reagan (Amherst: University of Massachusetts Press, 1999), *EH.Net* February 2001.

[2] For a detailed history of urban transportation planning in the U.S., see the most recent edition of Weiner's comprehensive history Edward Weiner, *Urban Transportation Planning in the United States: An Historical Overview*, <http://www.bts.gov/tmip/papers/history/utp/toc.htm>, 5th ed. (Washington, D.C.: U.S. Department of Transportation, September 1997).

Analytically, the next step was the development of a forecast of future demand. The simplest technique was to extrapolate from current demand using a growth factor. A more sophisticated approach involved developing travel demand models that used current data for calibration and then forecasted future demand on the basis of forecasts of land use and population.

During the 1950s, pioneering studies in a number of cities provided the foundation for the development of a common "best practice" for urban transportation planning. In 1948, San Juan, Puerto Rico, was the first of the major metropolitan studied, followed by Detroit (1953), Chicago and Washington, D.C. (1955), Baltimore (1957), Pittsburgh and Hartford (1958), and Philadelphia (1959). The Chicago Area Transportation Study (CATS) was the first to bring together all of the elements of what was to become the standard approach for the next several decades: data collection, forecasts, goal formulation, alternative network proposals, and testing and evaluation of alternative networks.[3]

The resulting model-based approach embraced five central analytic principles:

1. Trip generation: The principle that land use generated traffic and, by extension, that one could predict traffic on the basis of predicted land use.
2. Trip distribution: The principle that travel between two points was proportional to the attractiveness of the land uses at the destination and inversely proportional to the travel time (or other impedance) between them.
3. Modal split: The principle that users choose their mode of transportation between two points by evaluating the attractiveness of the modes available.
4. Traffic assignment: The principle that the specific route chosen for a trip between two points could be predicted by identifying the route with the shortest travel time (or other impedance).
5. Traffic service: The principle that network alternatives should be based on efficiency as indicated by the cost of travel.

Models based on this approach came to dominate metropolitan transportation planning in the U.S. The Bureau of Public Roads embraced the approach and provided funds to support its development and dissemination.

The Bureau consistently advocated regional transportation planning and integration with local highway systems. The 1962 highway act instituted a regional planning requirement for all cities larger than 50,000 in population—the so-called 3C planning requirement (continuing, comprehensive, and cooperative)—and set aside 1.5 percent of highway funds for planning. (The 1.5 percent had previously been optional.) By the legislated deadline of July 1, 1965, all 224 urbanized areas over 50,000 had an urban transportation planning process underway.[4]

The 3C requirement gave the Bureau authority to define what such a process should entail. The Bureau used that authority to establish standardized planning procedures, using a common bat-

---

[3] Edward Weiner, *Urban Transportation Planning in the U.S.: An Historical Overview*, DOT-T-93-02, rev. ed. (Washington, D.C.: U.S. Department of Transportation, November 1992), 31-32.
[4] Weiner, *Urban Transportation Planning in the U.S.*, 47.

tery of computer programs based on the analytical principles established in the major planning studies of the 1950s.[5]

This implementation decision had profound consequences: it established a mandated, standardized procedure for every metropolitan area in the country. The mandate entrenched in regulation four important provisions. First, it mandated the establishment of a planning body that encompassed the entire metropolitan region and provided it a source of funds to conduct transportation planning. The underlying assumption was that a regionwide body was the appropriate institutional arrangement for conducting transportation planning.

Second, it mandated a set of analytical procedures. These procedures may have been fairly well suited to the problem of deciding where to locate major highways. But it is now widely recognized that they have serious theoretical and practical flaws, and that they are not appropriate for many other important transportation planning problems, such as how to manage air quality and how to improve access by disadvantaged communities.

Third, the mandate established a set of evaluative criteria that gave priority to long-distance travel over local travel. The regionwide scale of the modeling exercise used a technique of dividing a region into zones and then conducting planning at the zone-to-zone level. Intrazone travel was beyond the scope of the regional planning undertaking. The result was that the longer distance travel that showed up in the regional planning models was anticipated and where possible accommodated. Shorter distance travel at the subzone level was beyond the scope of the required planning models.

Fourth, the criteria used to evaluate projects favored large facilities over small ones. The most important benefit of most projects was travel-time savings, usually valued at some fraction of the wage rate (sometimes 100 percent). In the typical static modeling framework, larger facilities are generally less expensive than small facilities on cost-per-unit of travel-time saved because of the economies of scale in highway construction. Bigger projects are cheaper to build on a per-trip basis than an equal amount of capacity configured in two or more smaller facilities. And since trip generation and distribution are traditionally assumed to be independent of facilities, larger facilities appear to provide more traffic service for the money.

The limitations of the four-step modeling approach are now widely acknowledged. One of the most serious limitation was the treatment of feedback between the various analytical stages—the "waterfall" problem. Each stage of the model is logically affected by subsequent stages, but due to computational and other limitations, feedback was at best only crudely reflected in the planning analysis, and more often not reflected at all. For example, it is clear that the construction of highways and other facilities affects land use. These models as usually implemented in the 1960s, however, did not capture that influence. Most applications of the models at that time treated future land use as fixed.

---

[5] Weiner, *Urban Transportation Planning in the U.S.*, 44, 46.

The development of so-called integrated urban models began in the 1970s to provide tools for systematically addressing land use and transportation interactions. Three land use models dominate current practice: (1) the Integrated Land Use Transportation Package (ITLUP) of S.H. Putman Associates, which consists of two main submodels, the Disaggregated Residential Allocation Model (DRAM) and the Employment Allocation Model (EMPAL); (2) MEPLAN, developed by Marcial Echenique and Partners; and (3) METROSIM, developed by Alex Anas. At present, however, many of the largest metropolitan planning agencies continue to use static, nonintegrated land use predictions for transportation planning purposes.[6]

Feedback is also clearly present in route choice, or the traffic assignment stage. In early implementations of the models, route choices between two points were based on the shortest path under uncongested conditions. Later implementations have utilized iterative and other techniques to ameliorate this problem, but it remains a difficult challenge today.

A second important limitation of the Chicago-style models is their nonbehavioral basis. The basic unit of analysis is the traffic zone, numbering several hundred in a large metropolitan area. Socioeconomic data are aggregated at the zonal level. But, of course, individuals—not zones—make transportation decisions based on individual socioeconomic characteristics.

Since the 1970s, efforts to improve transportation modeling have focused on improving feedback in the application of the models and on incorporating theoretically more appealing behavioral travel decision-making models into practice. The behavioral approach focuses on the individual rather than the aggregate traffic zone. If that decision process can be modeled—where, how, and when to travel given a residence and place of work and nonwork activities—then the predictive power of the models can be improved.

A fundamental problem with the behavioral approach, however, is that it fails to recognize the dynamic nature of human behavior, that is, that behavior changes over time. The behavior of individuals changes as they mature. And the behavior of individuals with similar socioeconomic characteristics changes as they learn how to improve their well-being through more, less, or different utilization of the transportation system. Firms and other institutions that use the transportation system also change their behavior as business processes improve and in response to the many changes that unfold in a competitive business and organizational environment.

A major innovation like a new transportation system precipitates many changes that are impossible to predict, especially at the outset of its deployment. How users use it determines its social and economic value, and hence the users' willingness to pay for it. Production, consumption, and distribution processes are highly complex and subject to constant innovation and search for competitive advantage. Some things are so valuable that users are willing to make big sacrifices to have them. Autonomous mobility seems to be one of them.

---

[6] Kazem Oryani and Britton Harris, *Review of Land Use Models and Recommended Model for DVRPC*, <www.bts.gov/other/tmippapers/landuse/compendium/dvrpc_toc.htm>, pub. no. 96008 (Philadelphia: Delaware Valley Regional Planning Commission, 1996).

More powerful computers have greatly improved the ability to implement iterative solutions to the feedback problem. Implementations of the four-step model are now available for desktop computers. Two or several iterations through the stages of the four-step model often produce convergence or near-convergence. Land use remains the most difficult part of the forecasting process, which is not surprising since actual land use is the result of complex and in many respects unpredictable actions by individual landowners, banks, local governing bodies, and a host of other institutions.

Beginning in the 1990s, efforts to improve transportation modeling took a new direction, spurred by a $25 million grant to Los Alamos National Laboratory in New Mexico. The Travel Model Improvement Program (TMIP), administered by the Federal Highway Administration, is an ambitious effort to improve existing models in the short to medium term, and over the long term to develop a new generation of travel models based on micro-simulations at the individual and household level.

Researchers in the transportation modeling community, many of whom had labored for two decades with little federal grant support, criticized the Los Alamos initiative because it gave so much support to researchers outside the traditional transportation establishment. But whatever its faults, TMIP brought a new analytical skill set to bear on transportation planning problems. Los Alamos scientists had enormous supercomputing capacity and extensive experience in developing models in particle physics. Their micro-simulation models model individual decisions about where, how, and when to travel on a day-to-day basis, including modeling the interactions of specific vehicles as they flow on to and congest traffic facilities throughout a metropolitan region. The objective is to use supercomputers to model every trip of every household in a metropolitan region on every transportation link, rather than simply modeling a sample of households.

TMIP developed a modeling suite, called TRANSIMS, and conducted a test case using a 25-square-mile portion of the Dallas/Fort Worth region consisting of 10,000 links and 200,000 travelers. The program has begun a second test case using a 25-square-mile portion of the Portland region consisting of 120,000 links and 1.5 million travelers, and has also announced a partnership with PricewaterhouseCoopers to make TRANSIMS commercially available.[7]

A problem with micro-simulation is getting data rich enough to estimate the models. Collecting real data can be expensive, and the data may be so detailed as to invade the privacy of its subjects. The ironic solution to this problem is to take aggregate data—measured, say, at the census block level—and then use statistical techniques to generate hypothetical individual data observations that can provide the necessary inputs to models. By design, the hypothetical data aggregate to match the original source data. Such synthetic data can improve insights into the sources and nature of transportation demand. It is not clear that they will improve the predictive accuracy of the models. This is somehow seen as providing better insights than working directly with the aggregates in traditional modeling.

---

[7] Nancy Ambrosiano, "Lab Gains Corporate Partner for Traffic Simulation," *Daily News Bulletin*, Los Alamos National Laboratory, October 5, 2000.

A difficulty is that theoretically attractive behavioral or micro-simulation models are often too complicated for decision makers who are not trained in the theory of the models. As a result, the models become the province of technical experts who often have limited political legitimacy.

Urban Travel Modeling and the Interstate Program

The limitations of the four-step process, now widely acknowledged, remain important in the context of current policy. Their uniform, nationwide application has given rise to a common set of problems for U.S. metropolitan regions. But the impact of the models can only be understood in conjunction with the programmatic features of the Interstate highway program.

Because the models failed to consider land use impacts adequately, they failed to recognize how quickly traffic would accumulate on newly opened facilities. Rather than providing capacity for growth over twenty years, as the models anticipated, new facilities often became congested soon after they opened.

Traffic growth that was more rapid than anticipated led highway departments to increase their estimates of future traffic for as-yet-unbuilt facilities. Already built facilities were expanded with additional lanes. Rather than recognizing that rapid traffic growth often reflected a spatial shifting of development from corridors and areas not yet served with new highways, many planners concluded that they had systematically underestimated the total demand for travel, and that all facilities needed to be scaled up accordingly.

The resultant scaling up of facility size and capacity dovetailed with the programmatic features of the Interstate highway program, which was financing much of the regional planning and construction of urban Interstates. The fiscal features of the Interstate program were a powerful incentive to build urban Interstate facilities first. At ten cents on the dollar, Interstates cost states and localities a small fraction of what it would cost to provide equivalent capacity off-Interstate. Facilities on other federal-aid systems cost states twenty-five to fifty cents on the dollar, or 250 to 500 percent more than Interstate capacity. And capacity built entirely off the federal-aid system cost 1,000 percent of equivalent Interstate capacity. Studies in the 1970s showed that states responded to these incentives by building their urban Interstate routes first.[8]

The result was that urban Interstates were built early, and to a very large scale. The increased forecasts indicated larger facilities to accommodate traffic for twenty years, and the Interstate program provided the funding necessary to build them. As it turned out, timing was everything. The early deployment of the Interstate routes, in the absence of new capacity elsewhere in metropolitan areas, led to intense development in Interstate corridors, with predictable consequences for Interstate traffic volumes. Increasing traffic volumes led to congestion and reduced levels of

---

[8] See, for example, Leonard Sherman, (Washington, D.C.: U.S. Department of Transportation, 1975) and Alan L. Porter and others, *Effects of Federal Transportation Funding Policies and Structures: Final Report*, tech. rept. no. DOT-P-10-80-09 (Washington, D.C.: U.S. Department of Transportation, 1979).

service on Interstates, thereby justifying their expansion with additional lanes, again at ten cents on the dollar. Bureau of Public Roads Chief Thomas MacDonald's prediction of 1940 that Interstates built in the absence of complementary facilities would distort land use and urban form played itself out in spades.

At the same time, local opposition to all highway construction, Interstate or otherwise, was growing. Concerns about the environment and neighborhood preservation gave rise to procedural and legislative changes that opened the planning process to a much broader constituency and essentially granted the power to veto or significantly delay highway construction projects to public interest groups.

National Environmental Policy Act

The passage of the National Environmental Policy Act of 1969 (NEPA) had important implications for regional transportation planning. NEPA introduced a new level of environmental review for transportation projects. It required the implementers of any project receiving federal funds—state highway departments in the case of the highway program—to conduct an environmental impact assessment and publish an environmental impact statement (EIS). The EIS had to include, among other features, an objective assessment of the environmental impacts of a project and an assessment of reasonable alternatives to the project. Moreover, the EIS was subject to approval by the federal government. The public could learn about projects through required public hearings, and public interest groups had legal standing to sue if they believed that the review process was procedurally or scientifically improper.[9] Environmental interest groups used NEPA to delay or stop many projects.

An important limitation of the NEPA process, however, was that it allowed projects to go forward even if they had acknowledged negative environmental impacts, as long as decision makers considered those impacts in the analysis. But NEPA did not provide absolute guidelines about which projects should be allowed to proceed. A second limitation was that its scope was limited to a specific project, such as a highway or corridor improvement. It did not allow consideration of the cumulative effects of multiple projects—that is, a regional transportation and land use plan.[10]

Hence, NEPA's impact on the metropolitan planning process was limited to the construction of specific projects within a plan. Cancellation and delay of specific projects as a result of NEPA forced plan modifications and the development of alternative improvements or policies to reflect cancelled projects. But the analytical process itself remained basically intact.

---

[9] "National Environmental Policy Act of 1969," *Stat.* 82 (1 January 1970): 852.
[10] Arnold M. Howitt and Alan Altshuler, "Controlling Auto Air Pollution" (Washington, D.C.: Brookings Institution, 1998).

Alternatives Analysis

Another important project-level planning initiative that left the regional planning process largely unaffected was the federal requirement for "alternatives analysis" for federally financed new urban transit projects (and major improvements). Civic enthusiasm for heavy rail transit systems as a solution—or important component—of regional transportation improvements began to mount in the 1960s and, after the oil shocks, in the 1970s. Many metropolitan areas sought federal funding for "new starts" of such systems. The federal government sought to encourage rational and systematic assessment of such projects as a condition of financing. In particular, it sought to ensure that localities considered a full range of alternative solutions, including less capital-intensive improvements such as light rail and enhanced bus service. Federal support was to be forthcoming for those projects where new starts were the most cost-effective.[11]

The analytical conflict revolved around what constituted cost-effectiveness. Rail advocates, in alliance with civic officials, often produced studies that reflected favorably on rail solutions. Skeptics argued that rail was often not cost-effective. While the conflict over the objectivity of alternatives analysis sparked a passionate debate in the transportation professional community, its impact was largely limited to the project level.[12] It did not have a major impact on the regional transportation planning process.

ISTEA

The metropolitan transportation planning and decision-making process was one of the prime targets of the reformers who influenced the passage of the Intermodal Surface Transportation Efficiency Act of 1991 (ISTEA).[13] Many calling for reform felt that urban and metropolitan interests were often subordinated to state departments of transportation, which were themselves often highway oriented.

The law called for several important changes (see Table 6-1). In general, ISTEA sought to elevate MPOs to leadership positions in their regions, as sources of objective, comprehensive analysis and planning. Plans were to be the result of broad-scale analyses of multiple scenarios and alternatives reflecting consideration of a comprehensive set of fifteen statutory planning factors, including air quality, travel demand, land use, efficiency, and equity.

---

[11] Weiner, *Urban Transportation Planning in the U.S.*, 128-30.
[12] U.S. Department of Transportation. Urban Mass Transportation Administration, *Urban Rail Transit Projects: Forecast versus Actual Ridership and Costs* (Washington, D.C., 1989); Martin Wachs, "Ethics and Advocacy in Forecasting for Public Policy," *Business and Professional Ethics Journal* 9 (1990): 141-57; Martin Wachs, "When Planners Lie with Numbers," *Journal of the American Planning Association* 55, no. 4 (1989): 476-79; John F. Kain, "Deception in Dallas: Strategic Misrepresentation in Rail Transit Promotion and Evaluation," *Journal of the American Planning Association* 55, no. 4 (1989): 476-79; Special issue on transportation forecasting, *TR News* 156 (September–October 1991).
[13] "Intermodal Surface Transportation Efficiency Act of 1991," P.L. 102-240, *Stat.* 105 (18 December 1991): 1958.

Table 6-1
Spectrum of Planning Practice

Aspect	Status Quo	ISTEA/CAAA Goals
General MPO role	Removed from major decisions	Broker, leader, consensus builder
Long-range plan	Single scenario	Alternative scenarios Multi-modal and inter-modal Focus on system performance Incorporates 15 factors
Links between plan and TIP	Not clearly established	Clearly established TIP strategic management tool
Fiscally constrained plan	No	Yes
Public role – Participation	Limited, e.g., hearings on plan /TIP	Actively encouraged; early and substantive.
– Representation	Limited	Broad public/private sector; citizens.

Source: William M. Lyons, "The FTA-FHWA MPO Reviews: Planning Practice under the Intermodal Surface Transportation Efficiency Act and the Clean Air Act Amendments," *Transportation Research Record* 1466 (1994): 23-30.
Note: MPO: Metropolitan Planning Organization; TIP: Transportation Improvement Program; ISTEA: Intermodal Surface Transportation Efficiency Act of 1991; CAAA: Clean Air Act Amendments.

ISTEA sought to give MPOs a much stronger role in decisions about long-range metropolitan transportation plans and the selection of particular projects for federal financial support. In urban areas the law placed authority for the development of the critically important transportation improvement programs (TIPs) in the hands of the MPO. TIPs contain the list of projects slated to receive federal support. The TIP was intended to become a strategic document that reflected the region's considered choice of development alternatives. It was not to be a laundry list of pet projects from parochial, uncoordinated local jurisdictions and operating agencies. Moreover, the TIP was to be "fiscally constrained," that is, within expected funding resources. Part and parcel of a heightened role for MPOs and the public was a reduced role for state departments of transportation, which had traditionally been the senior partners in selecting projects for federal support.

The law also called for greater public involvement in decision making. It sought to expand participation in the planning process to a broad range of stakeholders who would provide early and substantive inputs, rather than simply appearing at public hearings. Further, the law required the U.S. Department of Transportation (DOT) to certify the acceptability of a metropolitan area's planning process every three years as a condition for continued federal funding.[14]

---

[14] U.S. DOT issued interim guidance to implement the metropolitan planning changes outlined in ISTEA in April 1992, and final rules in October 1993. U.S. Department of Transportation, *Federal Register* 57 (April 23, 1992):

One difficulty with an increased role for MPOs was that many were not up to the task. Many had small professional staffs that lacked the resources to undertake significant planning studies. Objective prioritization of projects required planning data and analysis that many MPOs were unable to deliver in the short term. Many also lacked the political authority and independence to make and enforce tough choices about project priorities.

Between 1991 and 1996, the U.S. DOT sponsored a series of reviews of the transportation planning process in twenty-three major metropolitan areas. These reviews documented a number of areas where MPOs were falling short of DOT requirements for a certifiable planning process (see Table 6-1). Many MPOs were removed from major decisions. In planning often only one alternative scenario was considered. The projects comprising the TIP often bore little relationship to the realization of the plan, and participation in the planning process was often limited to agency and elected officials and organized interest groups. Southern California received a very positive evaluation, while Denver, among others, was judged to need substantial changes to come into compliance with the ISTEA guidelines.[15] To date, the U.S. DOT has not denied certification to any metropolitan area and has conditionally certified only a few.[16]

The impact of ISTEA on metropolitan transportation planning is hard to characterize. One of its leading lights, Lawrence Dahms, former head of one of the most progressive of the nation's MPOs, summed it up in a 1998 speech as "mostly resistance to the most challenging new concepts." State departments of transportation had "dug in" and resisted the heightened authority of the MPOs. And the principal reform group, the Surface Transportation Policy Project (STPP), had "chided everyone."[17] Another observer characterized it as a failure of perception and reality to connect.[18] The idea of the MPOs as the über-planning agency simply failed to recognize that control over the contents of the TIP still left substantial scope for control with the state departments of transportation and other operating agencies that controlled the awarding of contracts.

14943; U.S. Department of Transportation, "Statewide Planning; Metropolitan Planning," final rule, *Federal Register* 58, no. 207 (October 28, 1993): 58040-79. It issued an interim policy for public involvement in 1995. U.S. Federal Highway Administration and Federal Transit Administration, "Interim Policy and Questions and Answers on Public Involvement in Transportation Decisionmaking," notice, *Federal Register* 60 (January 27, 1995): 5463-68.

[15] William M. Lyons, "The FTA-FHWA MPO Reviews: Planning Practice under the Intermodal Surface Transportation Efficiency Act and the Clean Air Act Amendments," *Transportation Research Record* 1466 (1994): 23-30. No objective overview of these studies is available, although the U.S. DOT published an account of selected ISTEA "success stories" in metropolitan planning drawn from the studies in preparation for the reauthorization of the surface transportation program in 1997 and 1998. U.S. Department of Transportation, *Metropolitan Transportation Planning under ISTEA: The Shape of Things to Come* (Washington, D.C., 1997).

[16] Sheldon M. Edner, telephone interview (June 5, 1998).

[17] Lawrence Dahms, speaker, "Transportation —Show Me the Money," luncheon seminar, Center for Transportation Studies, Massachusetts Institute of Technology, March 13, 1998.

[18] Edner, telephone interview.

AIR QUALITY PLANNING

Public policy towards urban air quality has profoundly affected urban transportation planning. Public concern about air quality emerged in the 1960s, along with the rise of the American environmental movement. Congress first took action in 1955 with the passage of the Air Pollution Control Act, which it amended significantly in 1963. But it was the Clean Air Act Amendments of 1970 that established the major features of federal air quality policy. That policy, with minor amendments in 1977, governed until passage of a second set of major policy amendments in 1990.[19]

The key impact of the 1970s air quality policy on planning was that it required an assessment of how transportation investments would affect air quality. The 1990 amendments significantly strengthened the requirements for assessing the air quality impacts of regionwide plans and specific projects.

Period I: 1970-1990

The Clean Air Act Amendments of 1970 established three major policies that affected transportation and air quality. First, it empowered a newly created Environmental Protection Agency to develop and enforce air quality standards. Second, it required states to develop plans, called state implementation plans (SIPs), to achieve these standards within five years. In states that failed to submit an acceptable SIP, the 1970 amendments required EPA to submit a federal implementation plan (FIP). Third, it required auto manufacturers to reduce tailpipe emissions significantly.

In 1971, EPA promulgated the National Ambient Air Quality Standards (NAAQS), authorized in the 1970 amendments, and proposed regulations for the development of SIPs. SIPs typically governed not only mobile source emissions (such as those from motorized vehicles) but also stationary source emissions such as those from industrial sites. For metropolitan areas that would not meet the NAAQS in time, even with cleaner cars, SIPs were to contain transportation control plans (TCPs) to implement tailpipe emission reduction measures. Responsibility for the preparation of SIPs often fell outside the traditional urban transportation process and often did not involve agencies responsible for developing transportation plans.

Congress amended the Clean Air Act again in 1977. The amendments extended the deadlines for tailpipe emission reductions by two years and authorized EPA to waive them if they were not technologically feasible. The 1977 amendments required states and localities in nonconforming areas to submit revised SIPs to EPA by January 1, 1979, for approval by May 1, 1979.

---

[19] "Air Pollution Control Act," P.L. 84-159, Stat., 69 (14 July 1955), 322; "Clean Air Act," P.L. 88-206, *Stat.* 77 (17 December 1963): 392; "Clean Air Act Amendments of 1970," P.L. 91-604, *Stat.* 84 (31 December 1970): 1676-713; "Clean Air Act Amendments of 1977," *Stat.* 91 (7 August 1977): 685-796; "Clean Air Act Amendments of 1990," P.L. 101-549, *Stat.* 104 (15 November 1990): 2399.

The Evolution of Transportation Planning 123

The revised SIPs were to provide for attainment of the NAAQS by 1982, or in some extreme cases by 1987. Failure to comply could lead to the loss of federal-aid highway funds.

The 1970 and 1977 amendments had a significant impact on urban transportation planning. First, they dramatically reduced the planning horizon. SIPs prepared in 1978 for submission by January 1, 1979, were to provide for attainment by 1982, only four years later. The traditional horizon was twenty years.

A second major impact was on the type of projects included in plans. Most traditional transportation planning focused on long-term, large-scale, capital-intensive construction projects. Because SIPs and TCPs required actions to achieve the NAAQS in a much shorter time frame, they tended to emphasize low-cost and operational improvements. Third, planning techniques for a twenty-year horizon—flawed as they were for that purpose—were also quite unsuitable for air quality planning. The models could not represent transportation activity at a microscopic level. They had been developed for macro-scale planning. But clean air planning required micro scale analysis because the actual physical processes that affect air quality are transient over time and are highly susceptible to variations in local conditions. The disparity between the time horizon and spatial resolution appropriate for air quality planning and traditional metropolitan transportation planning led to disputes about the validity and utility of analytical models in debates over metropolitan transportation planning.

Nowhere have such disputes gone further than in a set of court cases in the early 1990s in the San Francisco Bay Area.[20] Two environmental groups brought federal lawsuits to force the local MPOs to use its models to analyze the long-term air quality impacts of several regional highway projects. In 1982, the MPO had adopted a contingency plan in the event it failed to attain the state ozone standard by 1987. Included in the contingency plan was a commitment to review the air quality impacts of specific highway projects and delay any with negative consequences. Such review has now come to be called "conformity" assessment, that is, the assessment of whether a particular project conforms air quality objectives in a plan.

When the region failed to reach the standard in 1987, the environmental groups sued the MPO to force it to conduct an acceptable analysis of the air quality impacts of several highway projects. The court case turned on the adequacy of the analysis. The environmental groups argued that the MPO's analytical approach overstated the air quality benefits of highway construction. The MPO's model included the air quality benefits of reduced congestion resulting from the projects. But it excluded the induced demand (i.e., traffic growth) that such projects often prompted, either by attracting development into improved corridors or fostering overall regional growth.

---

[20] *Citizens for a Better Environment v. Wilson and Sierra Club v. Metropolitan Transportation Commission* (Consolidated Cases), No. C89-2064THE (U.S. District Court for Northern District of California, 1991). This account draws on Greg Harvey and Elizabeth Deakin, "Air Quality and Transportation Planning: An Assessment of Recent Developments," in *Transportation and Air Quality*, part A, rept no. FHWA-PL-92-029, HPP-13/8-92(5M)E, 5, Searching for Solutions: A Policy Discussion Series (Washington, D.C.: U.S. Department of Transportation, Federal Highway Administration, August 1992); Mark Garrett and Martin Wachs, *Transportation Planning on Trial: The Clean Air Act and Travel Forecasting* (Thousand Oaks, Calif.: Sage Publications, 1996).

124 *FLEXIBLE URBAN TRANSPORTATION*

The MPO argued that the comprehensive modeling of induced demand was beyond the state of practice. Moreover, it argued, its models incorporated the analysis of induced demand as far as was practical (and indeed much further than most MPO analytical models). Theoretical models of the contribution of infrastructure improvements to regional growth, they argued, were not yet sufficient for inclusion in a conformity assessment.

The judge in the case appointed a special master, Martin Wachs, then a professor of planning at the University of California, Los Angeles. After reviewing the arguments of both sides, the judge ruled to accept the MPO's procedure. Further, he ruled that it was not reasonable at the time to expect the MPO to model the effect of infrastructure improvements on regional growth as part of its analysis.

The trial had several important implications for the role of transportation planning in air quality policy. First, it brought the technical merits of the transportation modeling process under judicial review and established legal precedents in federal court for adequate testing of conformity. The 1990 amendments to the Clean Air Act required conformity assessment in many MPOs, and the baseline laid down in the Bay Area cases may establish a de facto baseline for other MPOs, even though the Bay Area's procedures were considerably more sophisticated than those used in most other MPOs.

The second important impact of the trial was that MTC's analysis found that the model's fidelity fell far short of the factual requirements of the case in question. In attempting to estimate induced demand twenty years in the future, changes in travel times of as little as one minute induced significant changes in residential and job locations.[21] Virtually all observers agree that such sensitivity far exceeds the uncertainty in the models themselves.

Period II: The 1990s

The Clean Air Act Amendments of 1990 ushered in a new generation of air quality policies and, in conjunction with the reauthorization of the federal surface transportation programs a year later in ISTEA, an even greater level of integration between urban transportation planning and air quality planning. The 1990 amendments established several important policies that affected transportation planning. Two of the most important of these were its requirements for conformity assessment and its reliance for air quality improvements on so-called transportation control measures (TCMs).

The requirement for conformity assessment grew out of a requirement in the Clean Air Act Amendments of 1990 that no federal agency could approve or fund a project that did not conform to a state's plan for attaining the national ambient air quality standards. Projects had to be shown not to cause violations in and of themselves, nor to delay the attainment of the standards or intermediate benchmarks set forth in the plan.

---

[21] Garrett and Wachs, *Transportation Planning on Trial*, 123.

The conformity requirement had a very broad scope because many urban highway and transit projects receive some federal funds. Moreover, in implementing the regulations for conformity assessment, EPA elected to include not only projects directly receiving funds, but also any projects under the auspices of an agency that received federal funding, regardless of whether the project received direct funding.

The conformity rule is an ambitious attempt to integrate transportation facility planning and air quality. It governs all of the metropolitan regions that are now or have ever been persistently out of compliance with the national air quality standards (called non-attainment and maintenance areas, respectively). As of July 2000, 114 areas encompassing a population of 101 million people were in non-attainment of at least one criterion pollutant.[22] Moreover, in July 1997 EPA issued more stringent standards for ozone and for particulate matter of 10 microns or more (PM_{10}), and for the first time issued standards for smaller particulate matter, including particles in diesel engine emissions (those between 2.5 microns and 10 microns, or $PM_{2.5}$). The American Trucking Association and others joined to challenge the regulations, but the Supreme Court upheld them in 2001, and EPA plans to move forward with implementing regulations.[23] These initiatives will almost certainly increase the number of non-attainment areas that are subject to the conformity rule.

In regions subject to the conformity rule, virtually all transportation projects, regardless of funding source, are subject to review and approval by the federal government and to court challenge by any party that questions the data, methods, or procedures used to justify the project. Projects must be part of a federally approved plan for bringing the region's air quality into attainment with EPA's standards by a given deadline. This SIP must demonstrate how the region will achieve the air quality standards, taking into account the effect of all transportation construction and improvement projects in the region. Once the federal government approves the SIP, no project can be built that would cause violations of the standards or delay attainment of the standards or intermediate benchmarks.

The underlying premise of the conformity rule was that building highways could increase air pollution. The immediate effect of a highway improvement might be a reduction in congestion and hence a reduction in tailpipe emissions. But the longer term effects could well be the reverse. As travelers and developers adjusted to the availability of new highway capacity, traffic could increase and lead to greater, not less, air pollution.

The conformity rule sought to prevent such an eventuality by requiring projects to conform with federally approved state plans for achieving the national air quality standards. Thus, construction was not prohibited. But construction of new or expanded facilities had to be a part of a broader plan that would lead to attainment and maintenance of the air quality standards. As such, air quality trumps other considerations in the development of a region's transportation

---

[22] U.S. Environmental Protection Agency, "USA Air Quality Nonattainment Areas," http://www.epa.gov/airs/nonattn.html (July 31, 2000, access date: May 21, 2001).
[23] *Whitman et al. v. American Trucking Associations, Inc. et al.*, No. 991257 (U.S. Supreme Court, February 27, 2001).

system. Unless conformity can be demonstrated to the satisfaction of a critical audience of environmental interest groups, transportation facilities cannot be constructed or improved.

The other major feature of the 1990 amendments was its use of transportation control measures, or TCMs. The law identified sixteen TCMs that states or MPOs could include in their SIPs (see Table 6-2). The list includes a broad range of measures for potentially reducing tailpipe emissions, including promotion of transit, ridesharing, bicycling, and the removal of older vehicles from the vehicle fleet. The efficacy of many of these TCMs is questionable, however. It is difficult to demonstrate, for example, that promotion of bicycling has any measurable impact on a region's air quality. Moreover, some of the measures were deemed to be intrusive. Employer-based transportation management plans, for example, would have required employers to monitor the travel behavior of their employees and to pay penalties if too many employees drove alone to work.

Nonetheless, TCMs could be included by a region in its state SIP where they would count as Transportation Emissions Reduction Measures (TERMs). When evaluating a SIP for approval, EPA grants credit for certain air quality TERMs, depending on how they influence the regional air quality model. A carpooling program with a projected participation rate will give a nonattainment area a specific reduction in overall emissions in its SIP, for example.

The TCMs included in ISTEA were codifications of a strategy called transportation demand management, or TDM. TDM is a set of tools and techniques that began to emerge in the 1970s under the label of transportation systems management. Departments of transportation had traditionally solved congestion problems by expanding supply—that is, building new roads. TDM sought to address the demand side as an alternative to expanding supply, with all its environmental and community impacts.

The difficulty with TDM, however, is that it does not appear to be very effective in reducing traffic or reducing air pollution. A 1998 study of fifty-eight TDM projects in California found that, except for telecommunications projects (such as telecommuting), they were effective when compared to alternative fuel and fixed-route transit projects[24]—not a particularly ringing endorsement. In the meantime, the regulations are such that regions often adopt TDM because of the way the conformity credits work, not because they are effective in reducing traffic or air pollution.[25]

Even within the environmental community, the effectiveness of some TDM measures, such as high occupancy vehicle (HOV) lanes, are suspected of increasing traffic and pollution and reducing carpooling and mass transit usage because their effect is to increase overall transportation capacity.[26]

---

[24] C. Pansing, E. N.P Schreffler, and M. A. Sillings, "Comparative Evaluation of the Cost-Effectiveness of 58 Transportation Control Measures," *Transportation Research Record* 1641 (1998): 97-104.
[25] Jonathan L. Gifford and Danilo Pelletiere, "Study on Innovative Policies in the U.S.," report submitted to the Louis Berger Group, Inc. (March 2001).
[26] See for example, the advocacy piece Akos Szoboszlay, "HOV Lanes Cause Huge Solo Driver Increase on Montague Expressway," http://trainweb.com/mts/hov/hov-disinfo.html (July 1998, date accessed: May 18, 2001).

Table 6-2
Transportation Control Measures Authorized in the Clean Air Act Amendments of 1990

- Improved public transit
- Restricted roads or lanes for passenger buses or high occupancy vehicles
- Employer-based transportation management plans
- Trip-reduction ordinances
- Traffic flow improvement programs that achieve emission reductions
- Fringe and transportation corridor parking facilities serving multiple occupancy vehicle programs or transit service;
- Limited or restricted vehicle use in downtown areas or other areas of emission concentration particularly during periods of peak use
- High-occupancy, shared-ride services
- Portions of road surfaces or certain sections of the metropolitan area limited to the use of non-motorized vehicles or pedestrian use, both as to time and place
- Bicycle storage and other facilities, including bicycle lanes, for the convenience and protection of bicyclists, in both public and private areas
- Controlling of extended idling of vehicles
- Reduction of motor vehicle emissions caused by extreme cold start conditions
- Employer-sponsored programs to permit flexible work schedules
- Programs and ordinances to facilitate non-automobile travel and provision and utilization of mass transit, and to generally reduce the need for single-occupant vehicle travel as part of transportation planning and development efforts of a locality, including programs and ordinances applicable to new shopping centers, special events, and other centers of vehicle activity
- Programs for new construction and major reconstructions of paths, tracks or areas solely for the use by pedestrian or other non-motorized means of transportation when economically feasible and in the public interest
- Programs to encourage the voluntary removal from use and the marketplace of pre-1980 model year light duty vehicles and pre-1980 model light duty trucks

Source: Clean Air Act Amendments of 1990, P.L. 101-549, 104 Stat. 2399 (Nov. 15, 1990), §108 (f).

The fundamental dilemma is that the 1990 clean air act amendments relied heavily on reducing automobile travel, or slowing its growth, as a means for improving air quality. Most areas have not been successful in reducing travel demand. In fact, the booming economy of the mid- to late 1990s has led to steady increases in auto travel. Indeed, economic growth was been so strong that even many transit operators saw traffic increases, or a slowing of declines.

The conformity requirements and TCMs both focus on actions to reduce travel as a means for reducing the air quality impacts of travel. Another way to reduce the air quality impacts of road transport is to make travel cleaner by reducing tailpipe emissions. In fact, most of the improvement in urban air quality over the last three decades has resulted from cleaner cars, not from regulation of travel. However, emphasis on travel management has been driven in part by a concern that travel increases mitigate the effects of cleaner vehicles, slowing or reversing the improvements in urban air quality that the national policy seeks to attain.

Transportation Planning and the Control of Air Quality

The use of urban transportation planning as a mechanism for the control of air quality raises a number of problematic issues. First, the precision of the models as monitoring and prediction mechanisms is questionable. The urban transportation planning models in use in most metropolitan areas are simply incapable of providing evidence of sufficient precision to provide appropriate inputs for the air quality models. And the air quality models themselves are too imprecise for many of the types of monitoring that control strategies utilize.

A second problematic aspect of the use of transportation planning for air quality control is the limited efficacy of the policy instruments themselves: SIPs, TCMs, and conformity testing. SIPs are in no way "good faith" agreements to try to achieve common goals. SIPs have become adversarial legal documents, subject to challenge and enforcement by regulators and by environmental interest groups. As a result, the parties to SIPs—state departments of transportation, metropolitan planning organizations, and local governments—must be extremely cautious about the legal defensibility of their planning data, assumptions, models, analyses, and conclusions, and about the types of measures they include in the SIP.

The efficacy of many of the transportation control measures is also questionable. While many of the measures may be effective when implemented voluntarily by private companies or local jurisdictions, their efficacy under mandated conditions is limited. Employer-based ridesharing programs, for example, might cause a small reduction in travel demand if employers participate voluntarily. But legally mandated programs run the risk of precipitating resistance that could undermine the effectiveness of the program.

The conformity requirement has all the hallmarks of a Soviet-style command and control regulatory regime. Central agents set targets, and subordinate units must meet the targets or face sanctions. Debate about air quality policy focuses on procedural requirements: the measurement of pollution, modeling and forecasting techniques, intermediate pollutant "budgets," approved plans, lapsed plans, etc.

Preliminary results of a large-scale study of the conformity rule suggest that the best that can be said of the conformity policy's impact on air quality is that it has heightened awareness of air quality issues among transportation decision makers, and that it has fostered improved (or at least more extensive) communication between transportation planners and environmental regulators. That is, its effects have largely been of a "feel good" nature. It may have an increased effect in future years in rapidly growing non-attainment areas.[27]

Many clean air advocates have a broader policy agenda to prevent urban sprawl and preserve open space and agricultural land. But land use regulation is largely the province of local governments. SIPs require transportation plans, which depend on local land use plans, which in

---

[27] Howitt and Altshuler, "Controlling Auto Air Pollution"; Arnold M. Howitt and Elizabeth M. Moore, "The Conformity Assessment Project: Selected Findings," American Society of Civil Engineers presented at the Conference on Transportation, Land Use, and Air Quality, May 17-20, 1998, Portland, Oregon.

turn have important, if extremely complex, relationships with the extent and spatial distribution of housing and employment. Ultimately, control of land use through command and control regulation of air quality may be an extremely costly strategy. Moreover, to the extent it begins to become effective, it may foster a backlash in Congress that may ultimately impair the cause of clean air.

A third difficulty with using transportation planning to control air quality is the primacy that public policy has afforded to air quality. Under current policy, air quality trumps other social and economic objectives, including economic and community development. Whether such primacy is appropriate is ultimately a political judgment. Setting policy at a national level suggests an underlying premise that states and localities are incapable of reflecting the preferences of their constituents, or that such preferences ought not to govern. Most of the direct benefits of improving air quality accrue to populations local to the improvement. Nonlocal impacts include interregional transport of ozone and sulfur oxides and, to the extent they are significant, contributions to global warming. State and local institutions may well be better suited for establishing tradeoffs between economic and community development and air quality.

A final difficulty with holding transportation improvements hostage to air pollution control is the scientific basis for the air quality standards themselves. The problem is not that the criteria air pollutants cannot cause deleterious health effects. Instead, the emerging new science of exposure assessment holds that regulating emissions is "looking for pollution in all the wrong places."[28] The "right" place to start is with exposure, not with emissions. The key question is what are the greatest modes of human exposure to environmental toxins? Exposure-based regulation could lead to a very different set of priorities for reducing exposure, perhaps with dramatically different compliance costs.[29] If the greatest source of exposure to carbon monoxide is experienced by auto occupants while traveling in congestion, for example, the mechanisms for reducing that exposure include conventional solutions like reducing congestion, but they also include unconventional solutions like improved exterior air filtration.

> Hence, to protect public health best, the broad suite of environmental laws should be reexamined and judged by how effectively they reduce people's total exposure rather than by how they reduce total emissions. That effort would surely be substantial, both to recast a large body of legislation and to monitor how well the laws work to reduce exposure. But the payoff would be a dramatic reduction in health costs as well as an improvement in the economy and effectiveness of environmental regulation.[30]

Such assertions raise profound questions about the legitimacy of air quality policy itself. While exposure-based assessments of air pollution control strategies may validate aspects of current

---
[28] Wayne R. Ott, <http://www-stat.stanford.edu/~wayne> (n.d., accessed May 20, 2001).
[29] W. Ott, P. Switzer, and N. Willits, "Carbon-Monoxide Exposures Inside an Automobile Traveling on an Urban Arterial Highway," *Journal of the Air & Waste Management Association* 44, no. 8 (August 1994): 1010-18.
[30] Wayne R. Ott and John W. Roberts, "Everyday Exposure to Toxic Pollutants," *Scientific American* (February 1998): 86-91.

policy,[31] they also raise questions about the underlying strategies embodied in the Clean Air Act and reinforced with ISTEA and subsequent legislation.

Notwithstanding these limitations, it is important to recognize that the structure of air pollution regulation derives significantly from the traditional transportation planning community, which developed the foundational four-stage models in the first place. For more than a quarter century critics have questioned the large-scale modeling approach.[32] But it is so firmly entrenched that environmental interest groups have recognized that reform efforts must work within the framework defined by the traditional models.

The consequences of errors and limitations in the four-step model, however, have changed profoundly with their application in air quality policy. When the models were used to estimate the scale and scope of facilities in the 1960s and 1970s, the failure to incorporate feedback from facilities to land use had a significant impact on the shape of suburbanization—although, importantly, it did not cause suburbanization. Society has adapted to these impacts, and there is no real consensus about whether American-style suburbanization improves or diminishes quality of life. Suburbanization is highly energy- and land-intensive, but it affords an extraordinary range of choice for the many individuals who can and do choose it.

The use of the transportation models for SIP development and conformity assessment leads to errors with quite different implications. The ultimate lever is the ability to improve infrastructure as a means of fostering economic and community improvement and development. Giving prime status to air quality over the nature and extent of infrastructure improvement may shift economic activity among metropolitan areas, allowing those with favorable air quality to grow while those not in attainment of the air quality standards to stagnate. Alternatively, it may foster low-density growth outside of metropolitan areas. As agglomeration benefits decline with improvements in telecommunications, for better or for worse, increasing the difficulty of development within metropolitan areas may reinforce de-urbanization.

A better response to ill-founded transportation planning models ought not to be equally or more flawed air quality models. The response ought to be a decision-making process that reasonably incorporates the full range of considerations—equity, efficiency, human health, and land use—that bear on transportation policy decisions.

NATIONAL-SCALE PLANNING

Developments in metropolitan transportation planning have unfolded in a context of planning efforts at the national level. Early national planning efforts were limited primarily to studies of the nature and extent of the nation's roads. One of the earliest justifications for the federal high-

---

[31] E.g., Peter G. Flachsbart, "Long-Term Trends in United States Highway Emissions, Ambient Concentrations, and In-Vehicle Exposure to Carbon Monoxide in Traffic," *Journal of Exposure Analysis & Environmental Epidemiology* 5, no. 4 (October–December 1995): 473-95.
[32] Douglass B. Lee, Jr., "A Requiem for Large-Scale Models," *Journal of the American Institute of Planners* 39 (May 1973): 163-78.

way program, however, was the development of a system of highways. The need for a national system was underscored by the importance of highways in World War I. Railroads were not able to move all war materiel to the eastern ports and were temporarily nationalized. The military made extensive use of trucks for transporting materiel, inflicting considerable damage on roads unsuited for heavy vehicles. In 1922, General John Pershing produced for the first time a map of militarily strategic highways.

Throughout its history, military needs have played an important, albeit subordinate, role in the national highway program. National and civil defense were significant considerations in the planning and design of the Interstate highway system. Another important national highway planning consideration grew out of concerns about post–World War II economic conditions and a fear of returning to economic stagnation.[33]

Most national highway planning, however, has focused primarily on serving civilian purposes—that is, households and businesses. The methodologies for such studies have evolved considerably over the years. The earliest studies used an indicative planning approach to estimate highway "needs" using highway classification and engineering standards. Needs studies later gave way to so-called condition and performance studies. Recently, studies have also begun to use macroeconomic analyses of the relationship between highway investment and overall economic performance.

Highway Needs Studies

One of the basic tenets of early twentieth-century highway engineering was the importance of a "scientific" approach that incorporated proper facility location for access and reasonable grade, and road building materials and drainage for durability of the traveled way. A closely related design concept was that the design of a road should be well suited to its use. A logical corollary of these beliefs was that roads could be classified and appropriate design characteristics identified for each class.

According to this logic, a systematic, objective engineering study of roads could identify "deficiencies" in the road system, that is, facilities that were not appropriate to their use. Highway needs constituted the estimated funds required to correct these deficiencies. Highway needs studies were thus a form of indicative planning. Deficiencies relative to established norms of adequacy constituted a need.

The usefulness of highway classification was complicated from the outset by the existence of more than one administrative category of federal-aid highways. The 1921 highway act had designated two classes of highways to comprise the initial seven-percent federal-aid system: primary intercity highways, which comprised 3 percent of a state's total highway mileage, and secondary farm-to-market highways, which comprised 4 percent. In the 1930s, in response to

---

[33] See, for example, U.S. National Resources Planning Board, *Transportation and National Policy* (Washington, D.C.: U.S. Government Printing Office, 1942).

concerns about the Depression, Congress added another category to expand the number of road miles eligible for federal support. The initial 7-percent system was relabeled the primary system, and a new secondary system was created. Later in the 1930s, urban extensions of primary highways were added as another category. The legislative creation of the Interstate system in 1944 added still another category. Another category yet, an urban system, was added in the 1960s.

One difficulty with the proliferation of federal-aid systems was that each category of highways was authorized under a separate categorical grant program, often with different matching ratios. The inevitable result was that the classification of a particular highway began to be influenced by the type of funding available—a contravention of the whole notion of scientific highway engineering.

The creation of the Highway Trust Fund to finance the completion of the Interstate system added a further important dimension. A state's share of Interstate funding was based on the cost to complete its Interstate routes. States with many high-cost segments received a larger share than those with lower cost segments. This approach was a sharp departure from traditional allocation formulas, which relied on factors that could not be directly influenced by state highway officials, such as population, land area, and mileage of postal routes.

By the late 1960s, it was becoming increasingly apparent to highway administrators that the administratively defined highway systems used for distributing federal-aid funds were "inconsistent with any reasonable classification by function."[34] Highway classification was an essential component of rational planning, and so a major functional classification study was mounted in the early 1970s to classify highways according their function, irrespective of their federal-aid status.

The functional classification study defined a series of new highway systems for use in highway planning (see Table 6-3). Recognizing the differences in the structure of urban and rural highway systems, the classification system defined four broad functional classes that encompassed both urban and rural highways. These broad classes included functional subsystems that varied for rural and urban areas. Urban areas had more categories of principal arterials, while rural systems had more categories of collectors.

The significance of functional classification was that it provided guidance on the appropriate design standards for a highway: a highway's function should determine its design. Highways in the higher order systems, which were targeted towards longer distance (although largely still local) traffic generally merited higher design standards. Highways in the lower order systems merited lower design standards. With functional classification, technical criteria, rather than federal-aid status, could be brought to bear on issues of highway planning.

---

[34] U.S. Federal Highway Administration, *National Functional System Mileage and Travel Summary: From the 1976 National Highway Inventory and Performance Study* 2 (1977).

Table 6-3
Highway Classifications

Rural	Urban
Principal arterial system - Interstate - Other principal arterials	Principal arterial system - Interstate - Other freeways & expressways - Other principal arterials
Minor arterial system	Minor arterial street system
Collector system - Major collectors - Minor collectors	Collector street system
Local system	Local system

Source: U.S. Federal Highway Administration, *National Functional System Mileage and Travel Summary: From the 1976 National Highway Inventory and Performance Study* (1977).

The higher order functional systems tended, of course, to carry the lion's share of the nation's travel (see Table 6-4). The Interstate system (part of the principal arterial functional system), while it comprises only 1.2 percent of the mileage of the national highway system, carries 21.7 percent of its travel.

While there is a certain logical tidiness to functional classification, it is not a purely objective approach in practice. While making distinctions at the extremes is relatively straightforward, distinctions at the margins are difficult and, in some cases, arbitrarily dictated by administrative criteria, such as ceilings on the number of miles allowed within a state in a particular functional category. Furthermore, the length of the Interstate system was statutorily fixed.

The desire to apply "rational," objective criteria for facilities extended beyond the physical features of a facility, like its design speed, number of lanes, and signing and pavement marking conventions. Administrators also developed objective measures of levels of congestion called its "level of service" (LOS). LOS ranges from "A" ("very good") to "F" ("extremely congested"). Planning analyses at the facility level could then address capacity increases needed to bring the system to a given level of service. Such needs studies started initially as occasional reports in conjunction with major highway improvement initiatives, such as planning for the Interstate. Beginning in 1968, Congress required needs studies on a biennial basis.

By the early 1970s, however, the basing of highway program levels on needs studies based on classification and standards was coming apart. There was widespread concern that needs studies were more "wish lists" than they were objective statements of need. Furthermore, the cost of bringing the entire federal-aid system up to standard had grown beyond the capacity of the program to support it.

Table 6-4
Mileage, Lane Mileage, and Travel by Functional Highway System, 1999

Functional System	Road Length Miles	%	Lane-Length Lane Miles	%	Travel Distance Vehicle Miles (millions)	%
Rural						
- Interstate	32,974	0.8	134,200	0.2	260,204	9.7
- Other principal arterial	98,856	2.5	252,456	0.3	243,950	9.1
- Minor arterial	137,463	3.5	286,880	0.4	169,378	6.3
- Major collector	432,954	11.1	870,641	1.1	206,936	7.7
- Minor collector	271,690	6.9	543,378	0.7	57,617	2.1
- Local	2,097,926	53.6	4,194,494	5.1	125,545	4.7
Urban						
- Interstate	13,343	0.3	73,295	0.1	382,986	14.2
- Other freeways & expressways	9,125	0.2	41,496	0.1	171,563	6.4
- Other principal arterials	53,206	1.4	184,854	0.2	392,721	14.6
- Minor arterials	89,399	2.3	223,983	0.3	313,936	11.7
- Collector	88,008	2.2	186,348	0.2	131,613	4.9
- Local	592,978	15.1	1,185,953	1.5	234,886	8.7
Subtotal rural	3,071,181	78.4	6,282,049	7.7	1,063,630	39.5
Subtotal urban	846,059	21.6	1,895,929	2.3	1,627,705	60.5
Total	3,917,240	100.0	8,177,978	100.0	2,691,335	100.0

Source: U.S. Federal Highway Administration, *Highway Statistics, 1999* (Washington, D.C.), tables HM-20, HM-60, VM-2.

Condition and Performance Reports

The remedy to the ballooning of needs estimates was to refocus the reports on the condition and performance (C&P) of the highway system. The C&P reports utilized a more sophisticated system for monitoring the condition of highways. They sought to move away from the traditional needs study and focus more on the relationship between actual highway conditions, investment, and overall system performance. Congress extended the concept in 1991 to include a C&P report on transit.

The system for providing the data input to the C&P reports was the Highway Performance Monitoring System (HPMS), which consisted of a statistically valid sample of highway segments from each state containing measurements of pavement quality, congestion, utilization, costs, anticipated growth, and other data. On the basis of these data, the C&P reports estimated a number of funding benchmarks, such as:

- the future condition of the system if current funding levels are continued;

- the cost of maintaining the system in its current condition (the "maintain" scenario); and
- the cost of removing "deficiencies" from the system (the "improve" scenario).

The conditions and performance approach was a significant improvement over traditional needs studies. The problem with the conditions and performance reports, however, as with indicative planning approaches in general, is that they provide virtually no information about the economic value being created by the utilization of highway facilities. Condition assessments tend to focus on physical conditions, like pavement quality, and performance criteria tend to focus on the cost, especially program costs. But the condition and performance reports cannot address the question of whether the economy would be better off with more highway investment or less.

The 1997 C&P report introduced a significant improvement to the C&P methodology. Rather than evaluating investments solely on the basis of engineering criteria, the 1997 C&P utilized a benefit–cost criterion for highway investments, although it continued to utilize engineering criteria only for bridge assessments. Thus, the 1997 report provided economic criteria for program funding levels for the first time, including user costs (travel time, vehicle operating costs, and crashes) and system operator costs (maintenance costs and changes in salvage value). It acknowledged the importance of environmental costs, but was unable to include them because of the lack of consensus on the monetary value of air emissions.[35]

Another important limitation of the 1997 report was that it assumed substantial increases in demand for transit, even though "no American city has implemented any combination of policies consistent with the assumptions for transit demand used in this report."[36] It is a credit to the report that such limitations are openly acknowledged, but their presence nonetheless undermines the credibility of some of the conclusions.

Improvements to the C&P methodology continue apace. Indeed, the author participated in a panel in early 2001 to identify and prioritize improvements. Resource papers for the panel by senior transportation analysts provided candid assessments of the C&P methodology and suggestions for how it could be improved in the near and medium term.[37] While it retains significant limitations, it remains a centerpiece of the national planning and reauthorization process.

An important weakness that remains relates to the network effects of new facilities, which are discussed more completely in the next chapter. The current analytical paradigm tends to focus on the performance of existing networks. In the highway case, this is particularly evident in the

---

[35] U.S. Federal Highway Administration and Federal Transit Administration, *Condition and Performance: 1997 Status of the Nation's Surface Transportation System*, Report to Congress, <www.fta.dot.gov/library/reference/july98nr.html> (Washington, D.C., 1997), 58.
[36] U.S. Federal Highway Administration and Federal Transit Administration, *Condition and Performance: 1997 Status of the Nation's Surface Transportation System*, 52.
[37] Arlee Reno, "Potential Improvements—C&P/Reauthorization Bottom Line," draft, submitted to the National Cooperative Highway Research Program (February 12, 2001); and Kevin Heanue, "A Review of the Conditions & Performance Reporting Process," draft, submitted to the National Cooperative Highway Research Program (February 8, 2001); Michael Bronzini, "C&P Report Review," submitted to the National Cooperative Highway Research Program (February 5, 2001).

assessments of the Interstate system. The Interstate system carries a disproportionate share of the traffic—14 percent of urban traffic on 1.6 percent of urban mileage (see Table 6.3). Such a dominant role for the urban Interstates suggests that drivers find the Interstates preferable for a disproportionate share of their travel (at least as measured by vehicle mileage).

One way to interpret this is that the Interstates are so well located and well designed that they are preferred for a disproportionate share of travel. Interstates appear to be extremely cost-efficient. Another interpretation is that the non-Interstate highways are so unsuitable for most travel that drivers find the Interstates to be the only feasible travel alternative. Improvements to non-Interstate routes might actually reduce Interstate traffic. But such relationships between facilities at the network level are difficult to capture in the new C&P methodology.

A related difficulty arises from differences between system operators and system users over acceptable levels of system performance. For example, highway suppliers judge the level of service on a facility on the basis of, among other things, the level of congestion and delay during peak hours; more congestion is associated with reduced levels of service. Users, on the other hand, have quite a different assessment. While they clearly prefer less congestion, they are nonetheless willing to use facilities at times when they are congested. By whatever calculus they use, they judge themselves better off using the congested facility than not.

There are certainly situations where supplier norms and user preferences are largely congruent. Few users would use a bridge that had a known 10 percent chance of failing catastrophically unless there are no alternatives, and no respectable supplier would allow users on such a facility.[38] But when the consequences of failure are minor delay or slightly elevated risks of accident, the preferences of suppliers and users can diverge.

In a perfect market, of course, users and suppliers would eventually reach an equilibrium set of prices and service levels that satisfied both. But urban infrastructure users are not usually presented with a broad range of service–price combinations from which to choose, and suppliers are not usually engaged in a fee-for-service production operation. Rather, the range of choices available to users is sharply limited, and the revenues that support the supply function are significantly removed from the users' willingness to pay, either through non–facility-specific user fees or compulsory taxes. Under these conditions, subjective determinations of the appropriate level of service can be problematic.

The subjective nature of these criteria can lead to serious mismatches between the supply of facilities and the demand for them. System suppliers may strive to maintain technically based level-of-service criteria, while users, who operate under different preferences, seek to take advantage of facility supply where they can be better off using a facility than not.

---

[38] Cf. Arthur M. Wellington, *The Economic Theory of the Location of Railways: An Analysis of the Conditions Controlling the Laying Out of Railways to Effect the Most Judicious Expenditure of Capital*, 6th ed. (New York: Wiley & Sons, 1906), who makes this point about railway location vis-a-vis structures.

Economic Productivity Studies

Needs studies and condition and performance studies derive investment requirements at the national level by aggregating individual projects—a bottom-up approach. In 1989, an economist named David Aschauer caused something of a sensation in the transportation community by proposing that overall economic performance could be explained in part by the level of national investment in infrastructure.[39] Aschauer argued that the economic returns from investment in infrastructure in the American economy substantially exceeded the returns from private investment, and suggested that infrastructure investment be increased by substantial margins in order to improve the performance of the American economy. Critics of Aschauer faulted his statistical methods and claimed that correct estimates of the returns from public infrastructure investment would be much lower. His work has since spawned a plethora of econometric studies and critiques.[40]

Aschauer's work captured considerable attention in the political and professional community. Indeed, in 1990 the Transportation Research Board awarded him its highest honor, the Roy W. Crum Distinguished Service Award.

The work of Aschauer and his followers examines the productivity impacts of transportation and particularly highway investments. A simple example is constructing a bridge across a river. Assume that before the bridge is constructed there are two grocery warehouses, one on the west side and one east side of the river, which have responsibility for distributing groceries to their respective communities on either side of the river. The construction of the bridge suddenly allows those grocery warehouses to be consolidated. One of them can be closed, the other enlarged, or they can be closed and a new one constructed at a convenient location, allowing an increase in the scale of warehousing activities and overall reductions in inventory levels and inventory costs, providing benefits to the economy. Conventional analytical approaches fail to capture such productivity enhancements since these changes to the savings induced by being able to close a warehouse are not reflected in the traditional cost–benefit analysis of the traffic of the bridge crossing itself.

Macroeconomic analyses capture some of these external benefits and can provide a useful guide to the economically efficient level of investment. To this end, Federal Highway Administration commissioned a study to estimate the economic impact of highway investment in the post–World War II era. Returns from highway capital investment are roughly comparable to the returns from investment in private capital, about 11 percent, although the returns were much higher—on the order of 35 percent—during the 1950s and 1960s when the Interstate sys-

---

[39] David A. Aschauer, "Is Public Expenditure Productive?" *Journal of Monetary Economics* 23 (1989): 177-200.
[40] For review essays, see Edward M. Gramlich, "Infrastructure Investment: A Review Essay," *Journal of Economic Literature* 32 (September 1994). 1176-96; Marlon G. Boarnet, "Highways and Economic Productivity: Interpreting Recent Evidence," *Journal of Planning Literature* 11, no. 4 (May 1997): 476-86; Marlon G. Boarnet, "Infrastructure Services and the Productivity of Public Capital: The Case of Streets and Highways," *National Tax Journal* 50, no. 1 (1997): 39-57.

tem was being built.[41] But even this study cannot answer the question of whether continued expansion of the Interstate would have increased economic growth.

***

This chapter has provided a broad-brush overview of the evolution of American transportation planning in the twentieth century. A central thrust of the development of planning has been a search for objective criteria for assessing the appropriate location, design, and extent of transportation investments. Seeking after such a truth—meritorious as it is—is somewhat perilous, especially in a highly dynamic environment. The next chapter examines the nature of dynamic environments, and raises questions about techniques for planning and decision making in such environments.

---

[41] M. Ishaq Nadiri and Theofanis P. Mamuneas, *Contribution of Highway Capital Infrastructure to Industry and National Productivity Growth*, Prepared for the U.S. Federal Highway Administration, Office of Policy Development, work order no. BAT-94-008 (September 1996).

7

CHALLENGES TO THE NEOCLASSICAL ECONOMIC PARADIGM: COMPLEXITY, ADAPTATION, AND FLEXIBILITY

Chaos is the law of nature, order is the dream of man.
— Henry Adams

The search for an orderly process for the development of urban areas and their transportation systems has inspired and challenged urban planners and transportation planners for generations. Yet this search must confront the fact that the only completely orderly entities are static and inanimate. Disorderliness is a fact of life that poses problems for those who seek an orderly process for urban development. The disorderliness of a socioeconomic system like urban transportation is especially problematic.

There are (at least) two competing schools of thought about the nature of disorderly socioeconomic systems, both of which depart from the neoclassical economic model that has dominated most analyses of urban transportation systems. The increasing returns (or path dependence) school has emerged over the last decade to take issue with the neoclassical model's assumption that suppliers typically operate in the region of diminishing returns to scale and increasing average and marginal costs. They argue that technologies subject to increasing returns, which include many network technologies like road and transport systems, cause market failures that may in some cases justify government intervention.

The Austrian school challenges the neoclassical model's assumptions about the nature of market processes, the role of uncertainty in decision making, and the structure of capital. The Austrians view market processes as fiercely contentious and subject to profound uncertainty. What appear to be monopolies that are impossible to dislodge—say IBM's dominance in mainframe computing—turn out be subject to being overturned by new products and services that were not widely expected—say the emergence of desktop and network computing.

While the Austrians have questioned some of the more extravagant claims of the increasing returns school, both schools emphasize the importance of history in influencing economic outcomes. Whereas a neoclassical purist would emphasize how inexorable forces towards economic equilibrium shape individuals and institutions, the Austrian and increasing returns schools would emphasize how individuals and institutions shape and condition economic "progress."

Both schools also embrace the complexity of socio-technical systems. Indeed, increasing returns has been one of the foundational elements of a rather remarkable popular and scholarly interest (not to say a bandwagon) in the role of complexity and adaptation in the development of technical systems.

These alternative schools of thought on the nature of economic processes provide extremely useful insights into the development of urban transportation systems. For that reason, this chapter reviews the approach and worldview of each.

INCREASING RETURNS

Increasing returns has gained considerable popularity in recent years. Its exponents question the efficiency of markets in many domains, from industrial location to automotive propulsion systems to computer central processors and operating systems to typewriter and computer keyboards. Critics of increasing returns question both the properties ascribed to increasing returns systems and the public policy remedies they are used to justify. Because this is not a treatise on economics, the description here will emphasize the essential features of the increasing returns perspective, their limitations, and their applicability in urban transport.[1]

The increasing returns school takes some inspiration from recent research on chaos and complex adaptive systems in physics, mathematics, biology, and cognitive science. Complex adaptive systems are systems that operate far from equilibrium, steady state, or global optima, that are subject to improvement and anticipation by agents or decision-making units, and that adapt their behavior to changes in their surrounding environment.[2] Complex adaptive systems are of particular interest because most systems that social scientists study are complex and adaptive. Social, economic, technological, and environmental conditions are continuously in flux, continuously adjusting and adapting, in simultaneous interaction with their surroundings.

---

[1] This section is adapted from Jonathan L. Gifford, "Complexity, Adaptability and Flexibility in Infrastructure and Regional Development: Insights and Implications for Policy Analysis and Planning," in *Infrastructure, Economic Growth and Regional Development*, ed. D. F. Batten and C. Karlsson (Heidelberg: Springer-Verlag, 1996), 169-86. An excellent introduction to both sides of this debate may be found in the electronic discussion forum on path dependence begun in late 1996 and sponsored by Economic History Services, which is available at http://www.eh.net/lists/eh.res.php. Continuing discussion of the issues is also available on that site.

[2] This definition draws on John Holland, "Can There Be a Unified Theory of Complex Adaptive Systems?" paper presented at The Mind, the Brain and Complex Adaptive Systems (Fairfax, Va.: George Mason University, May 24-26, 1993). See also *Self-Organization and Dissipative Structures: Applications in the Physical and Social Sciences*, edited by W. C. Schieve and P. M. Allen (Austin: University of Texas Press, 1981); M. M. Waldrop, *Complexity: The Emerging Science at the Edge of Order and Chaos* (New York: Simon & Schuster, 1992); R. Lewin, *Complexity: Life at the Edge of Chaos* (New York: Macmillan, 1992).

Challenges to the Neoclassical Economic Paradigm 141

Neoclassical economics focuses on systems that tend toward equilibrium. Much of the contemporary world, however, appears not to be equilibrium tending at all.

The relevance of these insights to urban infrastructure systems is that most of the economic thought that has informed the development and understanding of infrastructure systems and their urban environment is firmly rooted in neoclassical economics. The models of von Thünen, for example, explain the relationship between transportation and land development. They were developed to explain the rather static world of an agrarian economy, with a single market center surrounded by an undifferentiated field of agricultural land.[3] To explain seventeenth- and eighteenth-century agrarian economies, of course, such models were useful because the assumption that the subject systems were fairly static and did exist in fairly stable equilibrium was at least plausible.

Conditions in contemporary industrialized countries are far from static, however. On the contrary, technological, economic, environmental, and social conditions are in a state of rapid flux. Applying decisions based on static analysis to a complex adaptive system may give rise to puzzling results, since the results predicted by the static model may often fail to materialize. For example, a famous paradox of dynamic network analysis, Braess's paradox, demonstrates that improvements to a single link of a network, which by static analysis would appear to be beneficial, may instead reduce overall network performance.[4]

Increasing returns economics squarely tackles the problem of disorderliness and unpredictability. Traditional economics addresses processes where returns to scale—that is, the benefits of expanding a particular activity—face diminishing returns. For example, the more widgets suppliers produce, the greater the competition in widget prices. More competitive widget prices squeeze marginal producers out of production; the result is an "efficient" equilibrium quantity and price of widgets being supplied on the market (see Figure 7-1).

The following discussion of increasing returns is in three parts. The first examines the sources of increasing returns. The second describes the properties of systems that are subject to increasing returns. The third discusses the policy dilemmas raised by the existence of systems subject to increasing returns. These issues are particularly salient in the area of technology standards, and hence much of the literature on increasing returns derives from concerns in that domain. But the issues and processes are relevant to urban transportation planning and development as well, as the discussion later in the chapter will show.

---

[3] Johann Heinrich von Thünen, *The Isolated State* (1826; reprint, Oxford: Pergamon, 1966).
[4] J. D. Murchland, "Braess' Paradox of Traffic Flow," *Transportation Research* 4 (1970): 391-94; Walter Knodel, *Graphentheoretische Methoden und Ihre Anwendungen* (Berlin: Springer Verlag, 1969), cited in L. J. LeBlanc, "An Algorithm for Discrete Network Design," *Transportation Science* 9 (1975): 183-99.

Figure 7-1
Increasing Returns

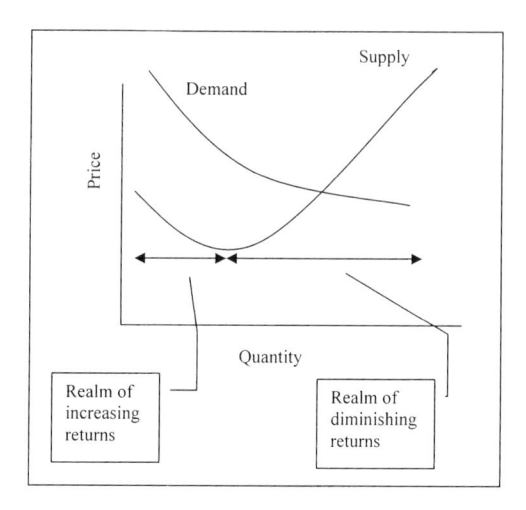

Sources of Increasing Returns

There are generally four sources of increasing returns: (1) coordination effects (also called network externalities), as in the telephone case cited above; (2) economies of scale and scope; (3) learning effects, "which act to improve products or lower their costs as their prevalence increases"; and (4) adaptive expectations, "where increased prevalence on the market enhances beliefs of further prevalence."[5]

Coordination effects

Coordination effects or network externalities refer to the technical interrelatedness of the components of a particular system. Technically interrelated components cannot be evaluated in isolation of the system of which they are a part, since their performance is contingent in some degree on the rest of the system. Congestion impacts are a common example in the urban transportation domain. Under congestion, the addition of one unit to a traffic stream reduces the system's overall performance. This would be a source of decreasing or diminishing returns. One of the purported justifications for the construction of the Denver International Airport, for example, was the network externalities of the runway separation at Stapleton International,

---

[5] This formulation of four sources is from W. Brian Arthur, "Self-Reinforcing Mechanisms in Economics," in *The Economy as an Evolving Complex System*, ed. P. W. Anderson, K. J. Arrow, and D. Pines (Redwood City, Calif.: Addison-Wesley, 1988), 10.

which it replaced. In low visibility conditions that required the closing of one runway, the effects rippled throughout the national air system resulted in delayed flights and missed connections.[6] Thunderstorms around O'Hare International Airport in Chicago, which can have effects across the national air system, are another example.

But coordination effects can also be positive. The imposition of a single truck width standard, for example, may lower overall costs for shippers and shift the split between trucking and other freight modes. Similarly, the expansion of a single bottleneck, say the new Denver International Airport, may improve the overall performance of a system, thereby reducing costs and benefiting other suppliers in that system.

A related coordination effect has to do with technologies that offer greater benefits as the number of users who make the same choice increases, that is, network externalities such as the choice to subscribe to a telephone service. The more people elect to subscribe to the telephone, the greater value there is to subscribe. If only one user subscribes, the system has no value at all. A similar example is the fax machine. Another is language.

A third type of coordination effect has to do with the cost of complementary products and services, which may be more readily or cheaply available if a large number of users elect to purchase the original product. Hence, the availability of a broad range of computer software for a particular operating system depends on the level of demand for that operating system. Similarly, the availability of a wide range of refueling and repair locations for automobiles is dependent on a large number of automobiles. This subgroup of coordination effects would also include traditional spatial agglomeration economies whereby there is an economic advantage for two firms to locate proximate to each other to take advantage of common external resources, like specialized law firms and labor pools.

Urban transportation clearly exhibits coordination effects. Economic activity involves the production, consumption, and distribution of material, energy, and information among households, firms, and other economic units. Infrastructure plays an essential role.

Coordination and interrelatedness in urban transportation are widespread. Households engage in distribution and consumption through the use of transportation systems, which requires accessibility, either by foot or other conveyance, to points of sale, be they strip shopping centers or neighborhood shops in a "walking city." Manufacturers distribute their products to points of sale using technical components that must be compatible. Distribution via rail cannot occur unless points of sale are accessible to rail heads. Distribution by truck cannot occur if streets are too narrow to allow their passage.

Network externalities are present as well. The benefits to a household of having a car increase as the number of desirable destinations that are car accessible increase, subject to transportation costs. The benefits to a retailer of being car accessible increase as the number of households

---

[6] There was some controversy about the need for the airport, based primarily on the belief that improvements in navigation technology could have allowed safe operation of both runways even under low visibility conditions.

owning cars increases, again subject to transportation costs. The benefits of living in a pedestrian friendly neighborhood increase with the number of shops locating in that neighborhood, subject to competition for space within walking distance. The benefits to a retailer of locating in a pedestrian friendly neighborhood increase with the number of people living within walking distance, subject to competition for space.

Complementary goods and services are also widely present in urban transportation. Most clearly, the more people that elect to own automobiles, the greater the diversity of auto-related services that are feasible, like service stations, tire outlets, etc.

Economies of scale and scope

A second important source of increasing returns are economies of scale and scope, that is, economies from larger scale or more diverse operations. The most familiar economies of scale are engineering economies, that is, the technical efficiencies that come from larger scale production. Often these take the form of large up-front fixed costs that allow lower variable costs of production. They might be attributable to, say, the fluid mechanical properties of production machinery. For example, the sidewall resistance in a pipe increases only in the first order of its diameter, while its carrying capacity increases in the second order, so that twice as large a pipe will have twice as much sidewall resistance but approaching four times as much carrying capacity.

Engineering economies are pervasive in urban transportation. For example, an investment in a lane of concrete between two points yields lower operating costs for moving vehicles between those points. Purchase of a vehicle (fixed cost) yields lower marginal costs of travel (excluding externalities). In the case of guideway (a highway lane or a steel track), congestion imposes decreasing returns beyond some level of traffic. With highways, the onset of congestion can sharply reduce the flow of traffic on a facility. With rail, capacity on a single line can increase to a maximum, but typically flow does not diminish due to congestion, delay simply rises.

One limitation of scale economies in guideways is the interference that can arise among additional units of capacity. For example, adding an additional highway lane provides additional capacity. But a breakdown in one lane can severely reduce the carrying capacity in adjacent lanes as drivers slow to exercise caution and "rubberneck." Similarly, accidents on one track can lead to slowdowns on adjacent tracks for safety reasons.

Scale economies also derive from production economies, which result from the lumpiness of capital investments and the possibility of redundancy in larger scale production operations. The use of one widget processing machine in a factory would leave that factory vulnerable to a 100 percent loss of capacity if that machine failed, whereas a larger factory that had four widget processors of equal reliability would be much less likely to lose 100 percent of its capacity.

An example of production economies in urban transportation is route redundancy in a dense network. A large provider, for example, can efficiently support multiple routes in a particular region, lessening the likelihood of total system shutdown due to the failure of a single link. For another example, a large freight carrier will on average have lower unit costs for any particular pick-up or delivery because of the likelihood of having other nearby pick-up or deliveries for any particular request.

Economies of scope occur when an increase in the range of goods and services produced brings about a decrease in average total cost. Large firms may be able to experience both economies of scale and economies of scope. Essentially, these are to human and organizational resources what engineering economics are to capital assets. For example, a small company might find it difficult to find a single employee who could act both as a personnel officer and an information systems expert, but not require a full-time person for each area of responsibility.

Economies of scope derive from being able to specialize particular parts of the production force. For a highway supplier, economies of scope would derive from being able to share maintenance garages across a wide range of facilities and service vehicle fleets. For a large fleet operator like a courier or taxi company, for example, such economies would derive from the ability to have a single service garage and single dispatch operation.

Learning by doing

Another important source of increasing returns derives from learning by doing.[7] A production organization gains experience in the process of production that allows it to produce the same product more cheaply and possibly to produce related or improved products more quickly or more cheaply than a firm just entering a production sector. For example, Japan is said to have built upon its early expertise in precision instruments to gain a foothold in the consumer electronics market and, subsequently, the integrated circuits that they contain.[8]

An important characteristic of increasing returns from learning is that they are not retroactive. The benefits of learning cannot be applied to decisions made before learning occurs. Thus, the learning process tends to lead down whatever initial path is chosen. Only failure along that path, or other strong conditions, can stimulate a return to an abandoned technological option.

At one level, life is a process of learning by doing. Which bakery has the freshest croissants? Which dry cleaner gives best service for the price? Which mode and route to work is most reliable? Most economical? Most secure? Anyone who has ever relocated to a new city knows that discovering alternatives and assessing them involves choice, experimentation, and learning that are irreversible and nonretroactive, in the sense that one cannot recapture expenditures

---

[7] Kenneth Arrow, "The Economic Implications of Learning by Doing," *Review of Economic Studies* 29 (1962): 155-73, cited in Arthur, "Self-Reinforcing Mechanisms in Economics"; Nathan Rosenberg, *Inside the Black Box: Technology and Economics* (Cambridge: Cambridge University Press, 1982).
[8] W. Brian Arthur, "Positive Feedbacks in the Economy," *Scientific American* (February 1990): 93.

made on wrong choices. At this level, households and firms are clearly subject to powerful increasing returns due to learning by doing. Habit, custom, and routine are all manifestations of learning by doing.

Certainly these learning-by-doing returns affect urban transportation. Routine routes to work and other destinations, perceptions about the comfort and security of using transit—all of these are instances of learning by doing.

In a similar way, agencies and governmental units are also creatures of habit, routine, and tradition. For sometimes rather obscure historical reasons, a particular policy or procedure is adopted. Even if that policy or procedure could never be justified starting from first principles today, it is necessary to show convincingly that an alternative will lead or is likely to lead to superior outcomes to justify departing from it.

This is not to say that households and firms are unwilling or incapable of adapting to changing circumstances. Quite the contrary! They seem to adapt with breathtaking speed when it serves their interests. But it does indicate that to induce a departure from routine, it is necessary for them to be convinced that the alternative is worth the trouble of trying it out.

One concern about learning by doing is that once one embarks upon a path, unchosen alternatives, which may have turned out to be preferable in the long run, are never revisited. In evaluating the merits of two competing technologies, in the presence of uncertainty about the relative merits of each, the cost reductions gained through learning by doing can make it unlikely that an untried technology will ever be pursued, even if its ultimate or long-term costs would be lower than the one that was tried.[9]

For example, some scholars have given this interpretation to the competition early in the century between petroleum-fired internal combustion engines and steam for automobile propulsion. If steam had prevailed and producers were as far down the learning curve with steam as they are now with internal combustion, society might be better off. Yet the costs of converting to steam today, given the uncertainties about its long-term costs, would be difficult to justify. While this interpretation of the competition seems far-fetched at best, it serves to illustrate how learning by doing could affect technical choice.[10]

---

[9] Robin Cowan, "Tortoises and Hares: Choice among Technologies of Unknown Merit," *The Economic Journal* 101 (July 1991): 801-14.

[10] For a conspiracy theory on this competition, see David R. Beasley, *The Suppression of the Automobile: Skullduggery at the Crossroads* (New York: Greenwood Press, 1988); see also, Robin Cowan, review of *The Suppression of the Automobile* (New York: Greenwood Press, 1988), by David R. Beasley, *The Journal of Economic Literature* 27 (December 1989): 1682-84; Clay McShane, "The Failure of the Steam Automobile," in *Down the Asphalt Path: The Automobile and the American City*, Columbia History of Urban Life (New York: Columbia University Press, 1994), chap. 5.

Adaptive expectations

The final important source of increasing returns is adaptive expectations, such as "bandwagon" effects. These refer to choice situations where a consumer presumption about the dominance of a particular alternative reinforces its adoption to the point that it becomes the dominant alternative. By some accounts, for example, the prevalence of the DOS operating system was largely attributable to the widespread assumption by consumers that IBM would become a de facto standard, thereby leading it to occur. In cases where increasing returns due to adaptive expectations are present, the sponsors of competing alternatives sometimes engage in strategic behavior to establish an installed base or early market presence that will lead consumers to believe that a particular format will eventually prevail.[11]

Adaptive expectations exert powerful influence in many elements of the urban transportation domain. One important aspect of adaptive expectation is the irreversibility or quasi-irreversibility of particular decisions such as construction of a road or location of a military base. While military base closing activities in the U.S. have provided a stark reminder that such large-scale decisions are not completely irreversible, nonetheless, households and firms often incorporate assumptions about the irreversibility of such actions into their decision making.

Properties of Increasing Returns Systems

Increasing returns give rise to four properties of interest: (1) multiple equilibria; (2) possible inefficiency, which may occur when one equilibrium is superior to another but fails to win out; (3) lock in, which occurs when one of the solutions or equilibria, once it becomes dominant, is difficult to exit; and (4) path dependence, whereby the "selection" of a particular equilibrium may depend on relatively small chance events.

Two of these—multiple equilibria and inefficiency—are relatively well acknowledged in the literature. Lock in and path dependence are less well known and have only recently begun to be discussed.

Multiple equilibria

One of the most interesting properties of increasing returns systems is that they can evolve into a number of steady states, equilibria, or configurations. This possibility means that the system does not tend toward an optimum state. Each of the steady states may have a different level of social welfare associated with it, and it is possible for a system to "settle down" into more than one possible configuration or trajectory. This is a significant departure from the single general equilibrium state associated with neoclassical economics.

---

[11] For a literature review, see Paul A. David and Shane Greenstein, "The Economics of Compatibility Standards: An Introduction to Recent Research," *The Economics of Innovation and Network Technologies* 1 (1990): 3-41.

148 *FLEXIBLE URBAN TRANSPORTATION*

Economists have long recognized the possibility of multiple equilibria.[12] International trade theory has acknowledged since the 1920s a problem of multiple equilibria, as for example when two countries can each produce in two industries, both of which have large set-up costs (e.g., aircraft and automobile). Given economies of scale, it may make sense for one country to produce the aircraft and the other the automobiles and to trade with each other to achieve their preferred consumption mixture. But which commodity is produced in which country is not cannot be determined by economic logic alone.

Spatial economics has recognized a similar problem in the concentration of industries in particular regions. When firms benefit from locating close to similar or complementary firms, the allocation of industries among regions may depend on historical accident, since several possible equilibrium allocations may exist.[13]

A similar problem exists in industrial organization. There may be several stable market shares of competing products, say between VHS and Betamax video cassettes or IBM-compatible and Macintosh microcomputers.

Multiple equilibria or steady states are certainly extant in urban infrastructure. Different cities demonstrate that patterns or structures or organizations of urban forms do exist and continue for long periods of time. While they are hardly static or in equilibrium, they do have stability. Indeed, a large literature documents the existence of significant variations in the form and transportation characteristics of metropolitan areas.[14] In simple terms, evidence of multiple steady states is present in discussions of the different types of cities—the walking city, the garden city, the automobile city, the edge city. And for centuries there has been conscious attention to the vernacular versus the designed city.

Possible inefficiency

Inefficiency may exist because one or more of the possible steady states may be optimal from a social welfare standpoint. Emergence of any but an optimal steady state would be inefficient, but, as noted above, there is no reason to expect this optimum to occur.

There is a considerable literature on the topic of the efficiencies of various urban forms, including a series of studies on the costs of sprawl.[15] Possible inefficiency may arise because it is possible to select something other than the most efficient steady state.

---

[12] The following examples are taken from Arthur, "Self-Reinforcing Mechanisms in Economics."
[13] Paul Krugman, "Increasing Returns and Economic Geography," *Journal of Political Economy* 99, no. 3 (1991): 483-99.
[14] See, e.g., Peter W. G. Newman and Jeffrey R. Kenworthy, *Cities and Automobile Dependence: A Sourcebook* (Brookfield, Vt.: Gower Technical, 1989).
[15] For a review of the literature, see James E. Frank, *The Costs of Alternative Development Patterns: A Review of the Literature* (Washington, D.C.: The Urban Land Institute, 1989); Robert W. Burchell and others, *The Costs of Sprawl—Revisited*, Transit Cooperative Research Program Report, no. 39 (Washington, D.C.: National Research Council, Transportation Research Board, 1998).

Lock in

Lock in refers to the difficulty of deviating or exiting from a particular steady state condition (or stable attractor in the terms of chaos theory) once it has been entered. Familiar examples abound. To return to the international trade and spatial economics example above, once these two hypothetical countries have settled which will supply autos and which will supply aircraft and have developed their industries, the energy and effort required to change that arrangement would be formidable indeed. They would be "locked in" to their particular steady state. Extraordinary effort would be required to finance and develop new plants and equipment in the upstart area, as well as to establish the labor force, supplier firms, and regulatory environment that would be conducive to that particular industry. Similarly, the effort required to convert from VHS to Betamax video formats if Betamax were judged to be superior on some grounds would be extraordinary. Sunk costs in VHS equipment among households are large, complementary institutions such as videocassette distribution and outlets are well established.

Concerns about lock in on an inferior alternative due to increasing returns must be tempered by the recognition that it is not inexorable. Increasing returns due to large fixed costs may be mitigated by the depreciation of those assets. If an alternative is superior enough, by this reasoning, it will afford someone an opportunity to switch and exploit that opportunity. Thus Japan's "lean production" process displaced "Fordism," and superior computer technology frequently renders older technology obsolete.[16] If an alternative is not sufficiently superior to generate such exploitation, by this reasoning it probably doesn't matter very much. Indeed, a spirited debate has taken place in the literature about the existence and importance of lock in and the related concept of path dependence (discussed further below).

On the other hand, some theoretical work has identified circumstances of "excess inertia," where the net benefits of switching can outweigh the net benefits of not switching, and yet because of the distribution of those benefits, no switch will occur. The reason is that no single agent may benefit sufficiently to start the bandwagon, so no bandwagon ever gets started.[17] (There is also a complementary concern over "excess inertia," whereby switching can occur even when it is not beneficial.)

The notion of lock in is appealing for understanding the nature of urban development processes.[18] Even casual observation suggests that once an area has undergone development, redevelopment that differs very much is a difficult enterprise. Indeed, in her study of the impact of great fires and disasters, Christine Rosen found that, to a striking extent, cities were rebuilt in much the same pattern as that which had existed before the disaster.[19] The reason lies partially in the tenure of land and the regulations and legal restrictions placed upon its use and

---

[16] My colleague Brien Benson helped me understand this point.
[17] Joseph Farrell and Garth Saloner, "Standardization, Compatibility and Innovation," *Rand Journal of Economics* 16, no. 1 (Spring 1985): 70-83.
[18] Josef W. Konvitz, *The Urban Millennium: The City-Building Process from the Early Middle Ages to the Present* (Carbondale: Southern Illinois University Press, 1985).
[19] Christine Meisner Rosen, *The Limits of Power: Great Fires and the Process of City Growth in America* (New York: Cambridge University Press, 1986).

treatment, which does not change as dramatically in the event of a disaster as the improvements built upon the land. Once it is platted and subdivided, it establishes a structure of ownership rights and responsibilities that are in many respects extremely rigid.[20]

History provides many instances of the magnitude of effort required to exit from locked in states. Urban renewal programs of the 1950s and 1960s were explicitly aimed at reassembling parcels of land with clear titles in order to allow the financing and construction of buildings with a substantially different character. In terms of increasing returns, urban renewal was an attempt to exit from a steady state that was judged to be suboptimal. The construction of the grand boulevards of Paris between 1852 and 1870 was the result of an extraordinary exercise of the power of the state to demolish large corridors in the interests of achieving a harmonious and technically efficient configuration of streets and vistas. Baron Haussmann could only execute such a plan with the power and authority of Napoleon III behind him.[21] In the U.S., Robert Moses exercised tremendous power to introduce new transportation technologies into the New York metropolitan area.[22]

Path dependence

Path dependence refers to the fact that selection among possible equilibria might well turn on relatively small-scale events and choices. There is a large and growing literature on this topic. Among the favorite cases is the so-called QWERTY keyboard and its path to market dominance. In an influential 1985 article, Paul David claimed that the QWERTY keyboard was developed in the 1870s to reduce the frequency of key clashes. It also allowed the product name, "Type Writer," to be spelled out using only the keys on one row. A competing configuration, "The Ideal," had keys composing 70 percent of English text on the "home" row of keys. Since then, another format, "Dvorzak," has also been suggested. But QWERTY's dominance has so far been difficult to dislodge.[23]

If QWERTY were objectively the "best" keyboard configuration, then its market dominance would be of only limited interest. But some argue that it is in fact objectively inferior to competing alternatives available at the time of its adoption and to the since-developed Dvorzak. Its dominance is attributable not to any technical superiority, according to this line of reasoning, but rather to historical "accidents" that occurred at the time of its development and marketing. Specifically, QWERTY's inventor, James Densmore, promoted his product through training programs that taught touch typing. As QWERTY-trained clerks proliferated, the manufacturer of the competing configuration started making QWERTY available as an option on its machines, and QWERTY emerged as the predominant configuration. Thus, the reasoning

---

[20] Hildegard Binder Johnson, *Order upon the Land: The U.S. Rectangular Land Survey and the Upper Mississippi Country*, The Andrew H. Clark Series in the Historical Geography of North America (New York: Oxford University Press, 1976).
[21] Howard Saalman, *Haussmann: Paris Transformed*, Planning and Cities (New York: Braziller, 1971).
[22] Robert A. Caro, *The Power Broker* (New York: Knopf, 1974).
[23] Paul A. David, "Clio and the Economics of QWERTY," *American Economic Review* 75, no. 2 (May 1985): 332-37.

goes, QWERTY's adoption is attributable to the merits of the training program, not the technology itself. Its selection was path dependent.

But David's view of QWERTY has been challenged both on specific historical facts, and on the underlying concept of path dependence.[24] This work has identified three forms of path dependence, termed first, second, and third degree. First-degree path dependence is that where initial conditions influence the long-term development of a technology, but do not lead to inefficiency. For example, the selection of a particular system for powering the machines in a plant may influence the plant for decades. But these long-term implications may be fully appreciated by the initial decision maker.

Second-degree path dependence takes into account imperfect information so that efficient decisions at one point in time may not appear to be efficient in retrospect. Thus, dependence on initial conditions and imperfect information lead to regret but not inefficiency.

Third-degree path dependence adds the element of "remediability," that is, sensitivity to initial conditions leads to inefficiency that was remediable but was not remedied. Only third-degree path dependency is inefficient and represents a market failure. But these critics claim that advocates of path dependence have failed to produce a single documented case of third-degree path dependence, and hence they question whether it can be that important to the path of economic history. One is tempted to agree and prefer the Austrian orientation outlined below.

Path dependence of the first and second degree is clearly present in urban transportation and land development. The events precipitating the location of a settlement or the particular layout of a city or street system are highly site specific and enigmatic in many instances. It is not that these were random events in the sense of being drawn from a probability distribution. They may well have been conscious and systematic from the standpoint of the decision maker who initiated the precipitating event. But the consequences of these decisions have extended far beyond the considered expectations of the decision makers who originally took them. Furthermore, the conditions have changed. The possibility frontier is now much different than it was. Sensitivity to initial conditions is clear. Information has certainly been imperfect. Regret is certainly present among some. But remediability is a very difficult case to prove.

THE AUSTRIAN SCHOOL

The Austrian School offers a much more fundamental challenge to the neoclassical model. Whereas the increasing returns school focuses on the possibility of market failure under conditions of increasing returns, the Austrians challenge the very question that the neoclassical school seeks to answer. Where the neoclassical school sees market competition as a *situation—*

---

[24] S. J. Liebowitz and Stephen E. Margolis, "The Fable of the Keys," *Journal of Law and Economics* 33 (April 1990): 1-26; S. J. Liebowitz and Stephen E. Margolis, "Path Dependence, Lock-in and History," *Journal of Law, Economics and Organization* (April 1995): 205-26; S. J. Liebowitz and Stephen E. Margolis, "Are Network Externalities a New Source of Market Failure?" *Research in Law and Economics* 16 (1994): 1-22.

that is, already stabilized—the Austrians see it as a *process*. The neoclassical school focuses on market competition as a means of establishing the correct price and quantity of a given set of goods and services. The Austrians assert that the market in no way offers up a given set of goods and services but instead consists of a process of discovering—through rivalry, experimentation, success, and failure—the goods, services, quantities, and prices that a society chooses to produce, distribute, and consume.[25]

The Austrian school originated in the nineteenth century and contributed many important economic concepts that are now part of the standard neoclassical paradigm. Utility, for example, originated as an Austrian concept, as did the concept of opportunity cost. Today's "neo-Austrians" focus on three central concepts that fall outside the neoclassical approach. The first is the notion of individual decision making as an act of choice under conditions of uncertainty. The second is the notion of markets and competition as a process of learning and discovery.[26] And the third is the notion of capital structure.

Choice under Uncertainty

One of the central tenets of the Austrian school is that individuals make decisions in an uncertain context, where information is imperfect and the identification of relevant alternatives is part of the decision itself.[27] There is always uncertainty, especially regarding the plans and actions of other people, but also about past, present, and especially future conditions. Information is thus partial (i.e., not comprehensive), contaminated (i.e., subject to error), and costly (i.e., not available without expenditure of time, energy, or other resources).[28] As a result, decision problems cannot be solved exactly and must be solved approximately.

The market is the mechanism for distributing information, through the price system, about the relative value of goods and services in the economy. Through the price system, individuals are able to judge the relative value of various alternative courses of actions without detailed knowledge of the preferences or opportunities available to others. In this way, the price system works as an information system.[29]

The economic problem that markets address is that no one person or organization knows the complete possibility frontier of the economy, that is, all possible combinations of price and quantity of all goods and services. In fact, "knowledge of time and place" is widely distributed throughout the economy. Each individual has some specialized knowledge of time and place that is not generally known. The cost of assembling that knowledge at a central decision-

---

[25] I am indebted to my colleague Don Lavoie for his helpful comments and recommendations on this subject.
[26] Israel M. Kirzner, "Austrian School of Economics," in *The New Palgrave: A Dictionary of Economics*, ed. John Eatwell, Murray Milgate, and Peter K. Newman (London, New York: Macmillan Press Ltd., Stockton Press, 1987), 145-51.
[27] Kirzner, "Austrian School of Economics."
[28] This definition from a distinctly non-Austrian source, J. F. Traub, "Complexity of Approximately Solved Problems," *Journal of Complexity* 1 (1985): 3-10.
[29] Friedrich A. von Hayek, "The Use of Knowledge in Society," *American Economic Review* 35 (1945): 519-30.

making point, if it were even possible, would be considerable. Moreover, because assembling data does not happen instantly, data available to a centralized decision maker can become outdated and inaccurate. Indeed, the whole concept of assembling knowledge at a central place may be questionable, even as a limiting assumption.[30]

A further important tenet of the Austrian school is that tastes and preferences are inherently subjective, that is, the utility derived from a particular activity will vary among individuals (termed "methodological subjectivism"). Thus, even if a central authority had perfect information about the possibility frontier of an economy, the point on that frontier that maximized individual satisfaction would be unknowable because of the variety of individual preferences that are not known to the central authority. Indeed, Hayek points out that the only point on the possibility frontier that society knows how to reach is that point reached by market competition.[31]

Competition as Learning

A second important holding of the Austrian school is that the market is a learning process—that is, a process, not a situation. The Austrian view is that competition in the neoclassical framework means exactly the opposite of what it means in lay terms. Neoclassical perfect competition means that "every market participant does exactly what everyone else is doing ... it is utterly pointless to try to achieve something in any way better than what is already being done by others, and ... in fact, it is not necessary to keep one's eyes open to what the others are doing at all." Under equilibrium, every market participant has fully recognized what is and is not possible and has properly anticipated the actions of other participants in the market. But at that point, it is not necessary to pay any further attention to the market. Competition ceases.[32]

Under the market process model of competition, entrepreneurs are constantly testing the market by offering products at a price in order to discover consumer preferences and willingness to pay. They may seek to compete by lowering the price of their product, by enhancing its features, or by informing consumers through promotions, advertising, or other selling activities.

Not only do suppliers learn under the Austrian model, so do consumers. As suppliers offer different products and services to the market, consumers test them and learn whether they are worth the price. So the production frontier and the consumer utility functions are intrinsically dynamic and not necessarily equilibrium-tending.

The orientation towards process rather than stationary equilibrium is one of the most distinctive features of the Austrian school. As a concept, it fits in nicely with a burgeoning body of

---
[30] Don Lavoie, *National Economic Planning: What Is Left?* (Cambridge, Mass.: Ballinger Pub. Co., 1985).
[31] Friedrich A. von Hayek, "Competition as a Discovery Procedure," in *New Studies in Philosophy, Politics, Economics and the History of Ideas* (Chicago: University of Chicago Press, 1978), 185.
[32] Israel M. Kirzner, *Competition and Entrepreneurship* (Chicago: University of Chicago Press, 1973), 89.

154 *FLEXIBLE URBAN TRANSPORTATION*

research and analysis on organizational learning and organizational change.[33] In very general terms, organizational learning refers to the capacity of people in organizations to assess the results of their efforts, rethink how they go about their tasks, and then use these new ideas to change established practices. That is, members of organizations learn over time from experience by making inferences about cause and effect from results and using those inferences as a basis for decisions about program development and agency management. In learning organizations, members become intelligence gatherers as they search out the connection between their work and its effects.[34] Advocates of organizational learning argue that the only way to gain competitive advantage in the future will be through an organization's ability to learn.[35]

Capital Structure

A third important element of the Austrian view has to do with the treatment of capital goods. Capital theory identifies two concepts of capital: financial capital, which is the fund of capital, and capital goods like buildings and factories. Both concepts are important to both the Austrian and neoclassical schools. The difference is that neoclassical discussions of capital tend to treat both as if they were like financial capital, that is, as if telephone wires and trucks could be added together to derive a total stock of capital goods, K. The illiquidity of capital assets—the fact that physical goods often cannot be easily transformed into cash—is treated in an ancillary manner as durability.

Austrians, on the other hand, tend to emphasize the heterogeneity of capital goods.[36] Capital takes two forms. The operating assets of a firm consist of its physical capital goods and the reserve funds needed to operate them. Securities are the titles to capital goods, and they represent the control of production and the recipients of payments. These operating assets are highly heterogeneous and cannot simply be added together because their economic value is highly dependent on how they are utilized by entrepreneurs. A clever entrepreneur may be able to make much more of a particular asset than one who is not so clever.[37]

---

[33] Peter M. Senge, *The Fifth Discipline: The Art and Practice of the Learning Organization* (New York: Doubleday/Currency, 1990); Peter M. Senge, "The Leader's New Work: Building Learning Organizations," *Sloan Management Review* (Fall 1990): 7-23; Chris Argyris, Robert Putnam, and Diana McLain Smith, *Action Science: Concepts, Methods, and Skills for Research and Intervention*, The Jossey-Bass Social and Behavioral Science Series (San Francisco: Jossey-Bass, 1985); Chris Argyris, *Overcoming Organizational Defenses: Facilitating Organizational Learning* (Boston: Allyn and Bacon, 1990); Chris Argyris, "Action Science and Organizational Learning," *Journal of Managerial Psychology* 10, no. 6 (1995): 20-26; Jonathan L. Gifford and Odd J. Stalebrink, "Remaking Transportation Organizations for the 21st Century: Learning Organizations and the Value of Consortia," *Transportation Research, Part A: Policy & Practice* 36, no. 7 (2002): 645-57..
[34] Julianne Mahler, "Influences of Organizational Culture on Learning in Public Agencies," *Journal of Public Administration Research and Theory* 7, no. 4 (October 1997): 519-40; Michael Marquardt, *Building the Learning Organization: A Systems Approach to Quantum Improvement and Global Success* (New York: McGraw-Hill, 1995).
[35] Senge, "The Leader's New Work"; Bob Garratt, *Creating a Learning Organization: A Guide to Leadership, Learning and Development* (Cambridge, England: Director Books, 1990).
[36] Don Lavoie, personal communication with the author (October 1, 1998).
[37] Ludwig M. Lachmann, *Capital and Its Structure* (London: London School of Economics and Political Science, University of London, G. Bell, 1956); 89.

To the Austrians, capital goods must continuously be redeployed to new uses that are consistent with ever-changing market conditions. Capital goods are like fossils, fossils of prior expectations about market opportunities and conditions. It is not surprising that capital goods often fail to live up to the expectations of their creators. What is central, however, is the *redeployment* of capital assets to other uses as market conditions continue to change.

> Malinvestment of capital may, in some cases, by providing external economies, becomes the starting-point of a process of development. A railway line built for the exploitation of some mineral resource may be a failure, but may nevertheless give rise to more intensive forms of agriculture on land adjacent to it by providing dairy farmers with transport for their produce. Such instances play a more important part in economic progress than is commonly realized. The ability to turn failure into success and to benefit from the discomfiture of others is the crucial test of true entrepreneurship. A progressive economy is not an economy in which no capital is ever lost, but an economy which can afford to lose capital because the productive opportunities revealed by the loss are vigorously exploited. Each investment is planned for a given environment, but as a cumulative result of sustained investment activity the environment changes. These changes in environment did not appear on the horizon of any of the entrepreneurial planners at the time when the plan was conceived. All that matters is that new plans which take account of the change in environment should be made forthwith and old plans adjusted accordingly. If this is done as fast as the new knowledge becomes available there will be no hitch in the concatenation of processes, of plan and action, which we call progress.[38]

Thus, each accounting period is not blessed with a sparkling new set of capital assets optimally designed to fit the needs, conditions, and opportunities during that accounting period. Rather, each period is serviced by a set of capital assets, some new, but most the legacy of prior decisions.

Complementarity of capital goods is central to the Austrian treatment of capital. Complementarity exists at two levels. Plan complementarity refers to the complementarity of goods within the plans of a particular entrepreneur and is the responsibility of the entrepreneur. Structural complementarity refers to the complementarity of capital goods in the economy and is, "if at all, brought about *indirectly* by the market, viz. by the interplay of mostly inconsistent entrepreneurial plans."[39]

The accumulation of capital goods also allows increasing specialization.

> As capital accumulates there takes place a "division of capital," a specialization of individual capital items, which enables us to resist the law of diminishing returns. As capital becomes more plentiful its accumulation does not take the form of multiplication of existing items, but that of a change in the composition of capital

---

[38] Lachmann, *Capital and Its Structure*, 17-18.
[39] Lachmann, *Capital and Its Structure*, 54.

combinations. Some items will not be increased at all while entirely new ones will appear on the stage....

The capital structure will thus change since the capital coefficients change, almost certainly towards a higher degree of *complexity*, i.e., more types of capital items will now be included in the combinations. The new items, which either did not exist or were not used before, will mostly be of an indivisible character. *Complementarity plus indivisibility* are the essence of the matter. It will not pay to install an indivisible capital good unless there are enough complementary capital goods to justify it. Until the quantity of goods in transit has reached a certain size it does not pay to build a railway. A poor society therefore often uses costlier (at the margin) means of transport than a wealthy one. The accumulation of capital does not merely provide us with the means to build power stations, it also provides us with enough factories to make them pay and enough coal to make them work. Economic progress thus requires a continuously changing composition of the social capital. The new indivisibilities account for the increasing returns.[40]

The result of such increasing specialization is the freeing of less specialized goods for other entrepreneurial purposes.

Technical progress means unexpected change. It is by no means the only form of unexpected change which entails modifications of the capital structure. In reality the capital structure is ever changing. *Every day the network of plans is torn, every day it is mended anew.* Plans have to be revised, new capital combinations are formed, and old combinations disintegrate. Without the often painful pressure of the forces of change there would be no progress in the economy; without the steady action of the entrepreneurs in specifying the uses of capital and modifying such decisions, as the forces of change unfold, a civilized economy could not survive at all.[41]

IMPLICATIONS FOR INFRASTRUCTURE PLANNING

These two schools of economic thought both raise profound questions about the appropriateness of the neoclassical economic model as a foundation for infrastructure planning. The increasing returns school suggests market failure under conditions of increasing returns, although government intervention is not necessarily indicated. And the conditions giving rise to increasing returns are clearly present in the infrastructure domain.

The Austrians also have much to contribute. More than many fields, infrastructure planning is intricately involved in the modification and continuous redeployment of long-lived capital goods. It is essential to orient infrastructure planning towards the redeployment of the "fossils" of prior expectations, toward mending the network of torn plans, toward exploiting the new productive opportunities afforded by the unexpected path of progress.

---

[40] Lachmann, *Capital and Its Structure*, 85.
[41] Lachmann, *Capital and Its Structure*, 126-27, emphasis added.

This is a very different orientation from that which informs most infrastructure planning. Most infrastructure planning grows out of the neoclassical school of comparative statics. The Austrian school recognizes that entrepreneurs modify their plans and redeploy their capital assets to take advantage of unexpected developments. In an entrepreneurial economy such as that of the U.S., refusal to acknowledge failed plans and to move quickly to redeploy assets to other uses can be profoundly damaging. It can lead to a growing disjunction between the demand and supply of capital infrastructure. This is inefficient for the economy, and damaging to its ability to afford social well-being to its citizens.

These two schools of thought both emphasize the need for adaptation and flexibility. The Austrians emphasize the need to promptly redeploy capital goods when reality fails to behave according to plan. Increasing returns underscore sensitivity to initial conditions and the unpredictability of long-term impacts.

Neither school of thought gives much comfort to the search for an orderly process for the development of urban areas and their transportation systems. For better or worse, the process is disorderly. Attempts to impose order through control are costly and perhaps of only limited effectiveness. Hence the dilemma of how best to plan and progress in a disorderly, unpredictable environment.

***

This chapter has reviewed two lines of thinking that challenge the dominant paradigm of neoclassical economic model: increasing returns processes and the Austrian school of economic thought. While quite different, both raise important questions about the utility of the neoclassical model's assumptions about economic equilibrium and the efficient allocation of resources that such equilibrium is thought to produce. Large capital assets like highways and transportation infrastructure clearly can have a significant effect on an economy's efficiency, but these two lines of thinking illustrate that the neoclassical model has significant limitations in understanding those impacts. The neoclassical model is central to traditional transportation planning. Thus, as the next chapter will argue, a new approach to transportation planning is badly needed.

8

THE NEED FOR A NEW APPROACH

To reach perfection one must adapt to circumstance.
—Baltasar Gracián, 1647

To judge people of the past, he sees no fog on their path. From his present which was their faraway future, their path looks perfectly clear to him, good visibility all the way. Looking back, he sees the path, he sees the people proceeding, he sees their mistakes, but not the fog.... but for us not to see the fog is to forget what man is.
—Milan Kundera, *Testaments Betrayed*, 1995

Urban transportation planning is at a critical juncture. Over the last half century, the planning and analytical framework governing highway and transit investments has become increasingly conflicted. The incumbent framework is grounded in the four-step model first developed in the 1950s and 1960s. This model has been instrumental in decision making, providing the basis for the establishment of a systematic framework and process for evaluating tradeoffs and investments.

From its inception, the incumbent model has precipitated controversy. There has been conflict around how to modify the incumbent model through procedural and analytical fixes. The steady increase in the number of required planning factors has stretched and strained the models.

The result is a process that is increasingly complex, protracted, conflict-ridden, and—most troubling—poorly equipped to address the mobility and accessibility requirements of a dynamic, modern society. As a result of the incumbent framework, decision makers are examining the wrong questions and learning the wrong lessons. For example, answers to the question, "Will the construction of a highway or transit project have a salutary or negative economic or environmental impact twenty years hence?" will be highly uncertain because of the unpredictability of economic and environmental factors.

The conditions of the 1950s that gave rise to the incumbent approach differ starkly from conditions today. The nation was embarking on the deployment of a major new transportation network, the Interstate system. Computing power was new and primitive. The very existence of the relationship between transportation and land use had only begun to be understood. By contrast, today while there may be disagreement about whether it makes sense to try to "build one's way out of congestion," no one is seriously contemplating the deployment of a major new transportation network, discussion of high-speed rail notwithstanding.

The era when master plans could be drawn up and implemented is over, at least for a time. Thus urban transportation planning should no longer be primarily about developing long-term master plans. The long-term, synoptic questions are no longer the right questions, and may never have been. To presume knowledge of future conditions and values is at best hubris, at worst an abdication of responsibility to recognize and prepare for the uncertainties that lie ahead. The question should be whether an improvement makes sense in the short- to medium-term in the specific location and situation where it is to be deployed. Transportation planning should enable the discovery of value, deploy capital assets for particular applications, monitor their use, misuse, or nonuse, redeploy or supplement those assets as their ability or inability to create value becomes evident, and refine and articulate the capital stock to serve emergent communities of users more efficiently. The time has come for a transportation planning suitable for a dynamic, creative society.

FOUR FALLACIES OF THE CURRENT APPROACH

The current long-term approach to transportation planning is subject to four important fallacies. The first is its focus on exogenous goals, the second its reliance on predictive modeling, the third its focus on efficiency, and finally its focus on public involvement.

The Exogenous Goal Fallacy

Engineering is often directed towards solving problems of the following form: "Given a set of goals, plan, design, build and operate x," where x is some engineered device or facility. Goals are exogenous to the problem. The limitation of this approach is that goals do not magically present themselves to problem solvers. Goals are not directly observable in the economy, nor are they self-evident. Goals only become apparent through the actions of households and firms. People often do not know exactly what they want and may not immediately know how to use a new product or service. Hence its use may evolve over time. And people's use of it may surprise the producer or service provider. What is observable is the combined product of goals and opportunities, what firms and households choose, subject to the alternatives available to them and their knowledge of those alternatives. Understanding the sources of these "revealed outcomes" poses an enormous challenge.[1]

One of the sources of this fallacy is the continuing quest within the field of engineering for design as a purely "scientific" enterprise, informed by hard technical principles and protected

---

[1] Michael G. McNally, "The Activity-Based Approach," in *Handbook of Transport Modelling*, ed. David A. Hensher and Kenneth J. Button, vol. 1, Handbooks in Transport (Amsterdam: Pergamon, 2000), 60.

from subjective, parochial politics, a kind of physics envy. Of course, technical knowledge has a critical role to play in the development of urban transportation infrastructure. No one wants to see facilities collapse or accidents take lives. But within the realm of the safe and technically feasible lies a vast array of options. Deciding how and when to select from that array—how to experiment and discover ways to create value—is more than a technical process—one that must engage the participation not only of transportation engineers, but also the broader social and economic community.

The Predictive Modeling Fallacy

A second and related fallacy is the transportation planners' predisposition to rely on medium- to long-range forecasts. Such prediction-based planning has tremendous logical appeal. Facilities often last for decades, and facility location decisions often exert influence for even longer. Knowledge of the future would allow the supply of facilities to serve future demand efficiently and effectively.

Knowledge of the future would also allow for fabulous stock market returns. But the appeal of such knowledge does not make obtaining it feasible or sound. Many long-term forecasts have been disappointing, in surface transportation and elsewhere. In the case of highways, demand growth has often dramatically outstripped predictions. Transit forecasts have many times spectacularly exceeded actual demand. Many forecasts are contingent on complementary actions that fail to materialize, such as the construction of other facilities or the conformance of land development with zoning designations.

Predictive modeling is one approach to understanding a complex phenomenon. A model suggests that a course of action will lead to a particular outcome. The actual response may deviate from the prediction. Analysis of those deviations allows refinements to the model, which then suggests a modified response to another course of action. In urban transportation, however, the laboratory is the real world. Continued application and refinement of the predictive model may eventually lead to predictions of acceptable accuracy. But the deviations from anticipated impacts often have significant social and economic implications that unfold over decades. The problem is compounded when a uniform planning process is imposed nationwide as it has been in the United States, for the laboratory then is not one city but hundreds.

Several problems arise from the predictive modeling fallacy. The first is an inclination to focus on aggregates rather than disaggregated data. Aggregates tend to behave better from a forecasting standpoint because they are easier to measure and extrapolate and do not require highly complex, data-intensive theoretical models. Yet aggregate measures are simply that: aggregations of individual decisions and activities. An extrapolation of the aggregate carries with it an assumption that on balance the constraints and opportunities that give rise to that pattern of behavior are stable for the duration of the forecast, which is twenty years in typical transportation planning. Such forecasts do not account for adaptive behavior and ignore the dynamic economic and social processes that take advantage of new opportunities. To assume such stability in an environment of rapid technological change and investment in new capital stock is naïve, if not irresponsible.

A second problem is that focus on aggregates may also bias decisions about the structure and density of a network. Urban transportation modeling has traditionally ignored local "intrazonal" trips and traffic on lower order facilities (collectors and local streets), as well as pedestrian, bicycle, and other nonmotorized transport. The models traditionally focus on forecasting travel on highways and arterial streets. The difficulty with this is that the aggregation of traffic largely on the sparseness of the network. Generally speaking, a sparse network of large facilities will have highly aggregated traffic. A dense "dendritic" network of small facilities, like a traditional street grid, will have much less aggregated traffic, and hence traffic that is much more difficult to predict. Emphasis on the predictable part of the network can bias investment towards larger scale facilities at the expense of lower order facilities.

Over the last three decades, efforts to improve transportation planning have focused on incorporating behavioral and activity-based models into practice. While such models are much more appealing from a theoretical standpoint, they do not take into account that behavior changes. "[T]he essence of economic growth is that changes over time are dominated by *changes* in technology, institutions and tastes."[2] Major new transportation systems like the Interstate induce changes that are impossible to predict over the long term, especially when deployed. Behaviorally based micro-simulation activity modeling faces an additional problem: obtaining data of sufficient richness without violating the privacy of individual households. Moreover, they are often too complicated for decision makers not trained in the theory of the models. As a result, the models become the province of technical experts, who often have limited political legitimacy.

Major new surface transportation systems are not being widely contemplated today, and as the Interstate-based highway system matures, it is plausible that long-term forecasts will become more reliable. Yet the societal response to the Interstate has been indelibly shaped by the early forecasts that were the foundation for the scale and scope of the system. The counterfactual question is what would have happened if twenty-year forecasts, which we now understand were naïve, had been foregone for a more incremental learning-by-doing approach? While interesting fodder for speculation, any answer must be as naïve as the predictions of the Interstate planners. Yet those predictions of half a century ago have today locked us in to a set of unappealing choices.

A third problem arising from the predictive modeling fallacy is a pernicious risk of reinforcing the status quo. Those with power and influence tend to dominate planning and decision making. So it is not surprising that the resulting plans often reinforce their beliefs and serve their interests. Those making these decisions may believe their judgments are socially beneficial. But plans and their realization exert a powerful influence on the shape and direction of devel-

---

[2] Simon Kuznets, *Modern Economic Growth: Rate, Structure and Spread* (New Haven and London: Yale University Press, 1966), paraphrased in Richard A. Easterlin, "A Vision Become Reality," review of *Modern Economic Growth: Rate, Structure and Spread*, by Simon Kuznets (New Haven and London: Yale University Press, 1966), *EH.Net* (2001).

opment, and a purportedly objective technical planning process can move perilously close to a self-fulfilling imposition of incumbent decision makers' beliefs and interests.

The Efficiency Fallacy

A third fallacy of the incumbent approach to transportation planning relates to its pursuit of efficiency. It is tempting for facility managers to focus on the efficiency of the assets under their management. For example, there is a highway system. Congestion means poor use of highway capacity. It is tempting to change society to make highways operate efficiently.

The bias towards facility efficiency is widespread in maturing infrastructure systems and their managerial bureaucracies and is apparent in allocation of energy, transportation, and water resources. Demand side management (i.e., using pricing or conservation measures) grows out of efforts to reduce demand for products and services.

Such demand-side measures have a place. But it is also important to learn from the demand for products and services and recognize that people are generating value from using them. For example, the time and energy it takes to make a trip in a congested network is a lower bound estimate of the value of that trip. That is valuable information. People value their trips so much that they are willing to sit on the freeway in congestion for a long time to complete that trip at peak hours. That provides a lot of information about how users value trips.

It is tempting to say that society should be organized to ensure the efficient use of transportation and other resources. The more appropriate objective is to arrange infrastructure to support an efficient society and economy. In a perfect world, we could achieve efficient allocation through variable pricing, low transaction costs, capital expansion, and modernization to accommodate willingness to pay. In reality, we are far from perfection. Nonetheless, the answer is not to change the world to make transportation operate efficiently, convenient as that might be for transportation professionals.

The Public Involvement Fallacy

The final fallacy relates to public involvement. Conventional wisdom, practicality, and the law all dictate public involvement in planning and design decisions. Many transportation reformers believe that current practice reflects values that are not legitimate and that public policy has been hijacked by the "road gang"—the automobile, highway, development, and petroleum industries— and does not reflect the core values of most citizens. Reformers hope that public participation can somehow reclaim ascendancy for dense, walkable development that preserves open space and reduces "bad" vehicle travel.

The road gang, in contrast, views reformers as elitists who seek to impose their values on the broader public by empowering their own special interest groups in planning and decision mak-

164 *FLEXIBLE URBAN TRANSPORTATION*

ing. In their view, the current auto-oriented society is not the outgrowth of a political–industrial complex that has hoodwinked the public into liking single-occupant vehicles and suburban McMansions, but rather the realization of society's underlying preferences.

Public involvement is an important tool for increasing input from customers. But it involves risk: creating a platform for extreme views and not getting information about mainstream users and stakeholders. Behavior is also a form of participation. Goals and objectives can be gleaned from careful observation of how people behave, how they use the various features and capabilities of capital goods. The dual challenge is to figure out how to get creative and useful public involvement through conventional forums like public meetings and hearings and how to modify decision processes to incorporate learning from what users are doing.

A REINVENTED TRANSPORTATION PLANNING PROCESS

A new process for transportation planning would avoid these fallacies. This process would recognize the tradeoffs and tensions between control, flexibility, and adaptive discovery.

Control

There is a way, at least theoretically, that the traditional approach can work: through control. Imposing control on an unpredictable environment can induce it to behave predictably, at least much of the time.

This relates to a fundamental principle of design, which is to impose order at one level to simplify processes at another level. Consider two people stranded on a volcanic desert island who wish to pass the time by dancing until they are rescued. They can either learn to dance on the uneven surface afforded by the volcanic rocks. Or they can create a dance floor: a small, level area that will allow them to dance normally.[3] (The problem of music is left to the interested reader.)

Highways are smooth platforms that allow rubber-tired vehicles to travel smoothly—a design decision. One could design automobiles to traverse unimproved surfaces, much as a tank, tractor, or lunar landing vehicle might.

The fundamental question, then, is whether predictability is important enough to justify the social and economic costs of control. What are the options? Does urban transportation planning *require* control? Or better, what *level* of control is necessary for urban transportation planning?

Urban transportation is already subject to all kinds of controls: speed limits, driving-on-the-right conventions, and requirements for licenses, insurance, seat belts, and tailpipe emissions,

---
[3] I thank Börje Johansson for this analogy.

to name just a few. Yet one fundamental characteristic of automotive transportation is that it is not restricted as to time and place. With some exceptions, one can jump in the car at any time and journey to a chosen destination. Hence the question: is controlling trip-making behavior desirable? Do the ends justify the means? How effective would it be? And what are the alternatives?

Flexibility

Sometimes control is not an option. Under evolution, species either adapt or perish. Businesses adapt to changing conditions or go bankrupt. Surfers adapt to waves or wipe out. Not that control is undesirable—quite the contrary. To the business facing bankruptcy or a species facing extinction, control over hostile forces would be most welcome. But it may not be an option.

Flexibility and adaptation are complements to control. Flexibility is the readiness to adapt to circumstance. It is useful in the face of uncertainty about an appropriate course of action. With complete information and no uncertainty about past, present, or future conditions, flexibility is superfluous. It is only useful in the presence of uncertainty and the expectation of receiving better information in the future.

Flexibility and adaptation are not absolute strategies. They may be part of a portfolio of responses. In war fighting, one's enemy may not behave predictably. Success comes not only from predicting but also from being prepared for the unexpected. Indeed, the military has internalized the scientific paradigm by improving its ability to respond, and not relying exclusively on predictive modeling. For example, the 1997 Quadrennial Defense Review characterized the core of U.S. defense doctrine as "shape, respond and prepare."[4]

The concept of flexibility has attracted considerable attention in recent years in a broad range of disciplines, including regional science, economics, operations research, planning, and manufacturing. It has also long been a concern in military strategy (see Table 8-1; the chapter appendix contains a brief bibliography on the concept of flexibility). This interest in flexibility reflects in part an increasing awareness of the importance of uncertainty and risk in planning and decision making in general.

The plain-language meaning of flexibility is pliable or responsive to change. Flexible machines, then, are those that can produce new products or produce old products in new ways, like robots and computerized numerically controlled (CNC) machines. Flexible manufacturing systems (FMS) are integrated combinations of flexible machines that create small-scale "manufacturing cells" of computer-integrated manufacturing (CIM). FMS may integrate phases of the manufacturing process, such as design (using computer-aided design, CAD), production, or distribution. Flexible specialization and integration is a product strategy of small batch production of specialized products that may be economical where there is a large range of specialized products

---

[4] U.S. Department of Defense, *Report of the Quadrennial Defense Review* (Washington, D.C., May 1997).

Table 8-1
Concepts of Flexibility

Plain language	1.a. Capable of being bent or flexed; pliable. b. Capable of being bent repeatedly without injury or damage. 2. Susceptible to influence or persuasion; tractable. 3. Responsive to change; adaptable: a flexible schedule. (American Heritage, 3rd ed.)
Context sensitive design	Flexible application of highway design standards to incorporate sensitivity to the surrounding community or natural context. U.S. Federal Highway Administration 1997.
Flexible machines	Robots and computerized numerically controlled (CNC) machine tools that can produce new products or old products produced in new ways. Reduces the need to scrap fixed capital in order to product new products. Gertler 1988.
Flexible manufacturing systems (FMS)	An integrated combination of flexible machines to create small-scale "manufacturing cells" or computer integrated manufacturing (CIM). FMS may integrate phases of the production process, including design (using computer aided design, CAD), manufacture or distribution. Examples of distribution in FMS include "just in time" production systems. Gertler 1988.
Flexible specialization and integration	A product strategy of small batch production of specialized products, which may be economical where there is a large range of specialized products over which to amortize the cost of FMS equipment, which is usually more expensive than conventional equipment. Gertler 1988.
Flexible accumulation	1. Sociologists and political scientists use the term to describe a national economic strategy using new technologies, products and services and "skill-flexible" labor. 2. Geographers use the term to refer to a "sea change" in the way capitalism works. "It rests on a startling flexibility [in] labour processes, labour markets, products, and patterns of consumption, ... characterized by the emergence of ... new sectors of production, new ways of providing financial and business services, new markets, and ... greatly intensified rates of commercial, technology and organizational innovation." David Harvey, quoted in Gertler 1988.
Flexible regions	Regions with diversified industrial bases can adapt to changing economic conditions, whereas regions dependent on a single industrial sector can suffer when that industry is in a period of decline. Malecki 1991.
Military	One tries to maintain tactical and strategic flexibility on the battlefield by having logistics operations that can support a broad range of tactical and strategic options. There is a logistic element, and intelligence element. "Strategy must have at its disposal a variety of weapons and forces so that the particular combination most suitable to the situation, as it actually arises, may be quickly formed and swiftly and decisively employed in an appropriate manner." Eccles 1965.
Anthropology	Behavioral adaptation in response to changing environmental circumstances. Flexibility is the counterpart to adaptation. Without it, no adaptation; without adaptation, death or decline.
Biology	Genetic and developmental adaptations to changing environmental circumstances.
Management	Flexibility is a response to uncertainty about future conditions and the expectation that better information will be available in the future. Organizations may adopt incremental approaches to problems when there is uncertainty or conflict about goals or objectives. Organizations may also strive to be flexible in their response to problems and to be "learning organizations." Quinn 1978, 1989.
Land use and urban planning	Future land uses are difficult to predict and control, so flexibility is important for the economic vitality of a region. Planned communities can be sterile and lack the vitality of vernacular, unplanned communities. Malecki 1991.
Austrian economics	Only decentralized decision making (through markets) can make best use of the information available in the economy. Capital goods are fossils of plans based on partially or wholly unrealized expectations. Prediction is inherently imperfect. What is important is that capital goods be redeployed as quickly as new information becomes available.
Transaction cost economics	The decision to make or buy a good or service depends on the "specificity" of the good or service. If the good or service can be readily described and captured in a contractual document (using industry standards, specifications or plain language), it may be efficiently produced and traded in markets. If it is unique or has multiple intangible elements, it may be more efficiently produced internally.

Note: References drawn from bibliography in chapter appendix.

over which to amortize the cost of FMS equipment, which is usually more expensive than conventional equipment.[5]

The concept of flexible accumulation, a national economic strategy using new technologies, products, and services and "skill-flexible" labor, has been examined within the discipline of sociology. Geographers have used the same term to describe a "sea change" in the functioning of capitalism arising out of a "startling flexibility [in] labour processes, labour markets, products, and patterns of consumption ... characterized by the emergence of ... new sectors of production, new ways of providing financial and business services, new markets, and ... greatly intensified rates of commercial, technology and organizational innovation."[6] The concept of flexibility also has appeal in a range of other disciplines, from anthropology to biology to management to land use and urban planning.

Control and Flexibility

The major U.S. transportation legislation of the 1990s (Clean Air Act Amendments of 1990, the Intermodal Surface Transportation Efficiency Act of 1991, ISTEA, the National Highway System Designation Act of 1995, and the Transportation Equity Act for the 21st Century, TEA-21) resolved urban transportation policy around two conflicting agendas. On one side, environmental interest groups strove for more control over the air quality, land use, energy, and climatic impacts of urban development and did so by using transportation policy levers. On the other, the traditional transportation community was concerned about increased congestion and its threat to continued productivity improvements, economic growth, competitiveness, and quality of life.

The challenge for urban transportation planning today is helping society discover the best combination of control and flexibility, recognizing that the conflict between environmental integrity and economic vitality are not fundamentally about technical questions. There are technical components to these questions, to be sure, but the conflicts themselves are about values.

Figure 8-1 illustrates this. On the vertical axis is the level of agreement about the impact of a particular course of action. On the horizontal access is agreement about goals. In cell 1, where there is agreement about both goals and means, problem solving is fairly straightforward and is based on known techniques to achieve agreed upon ends. Cell 2 depicts agreement about goals but uncertainty about means. The impacts of particular courses of actions are not known with certainty, but the evaluation of outcomes is straightforward since there is goal agreement. Cell 3 situations are those where there are disagreements about goals but relative certainty about impacts. Cell 4 situations have neither agreement about goals nor certainty about impacts.[7]

---

[5] Meric S. Gertler, "The Limits to Flexibility: Comments on the Post-Fordist Vision of Production and Its Geography," *Transactions of the Institute of British Geographers*, n.s., 13 (1988): 419-32.
[6] David Harvey, quoted in ibid.
[7] I thank my colleague Louise White for this framework.

168 FLEXIBLE URBAN TRANSPORTATION

Engineering, technical efficiency, and Progressive ideology are typically most comfortable in cells 1 and 2, that is, in situations where there is agreement about goals. Indeed, the urban transportation process was originally developed as if it were a cell 2 situation: agreement on goals but uncertainty about impacts.

Today, the situation is much different. There is considerable disagreement about the relative importance of the goals of environmental integrity and economic vitality. Public policy is attempting to work in cell 3: to negotiate as if there were relative certainty about the impacts. However, long-term impacts are not known and cannot be known. The "correct" combination of environmental integrity and economic vitality can only be discovered through an adaptive, flexible process. The correct path will emerge in retrospect. But it is necessary to proceed today.

Figure 8-1
Framework of Analyzing Goals and Means

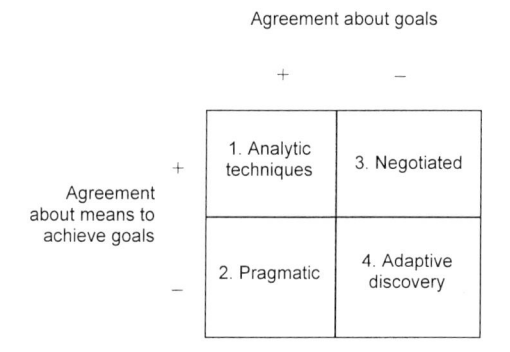

Agreement about goals

	+	−
+	1. Analytic techniques	3. Negotiated
−	2. Pragmatic	4. Adaptive discovery

Agreement about means to achieve goals

Source: Adapted from Russell Stout, "Management, Control and Decision," in *Management or Control* (Bloomington: Indiana University Press, 1980), 101.

A negotiated, long-term cell 3 solution risks locking society into the assumptions and beliefs of those with power and influence today. It risks foreclosing options that may become preferable as better information becomes available, as better technology develops, and as society learns more about how it values the environment and economy.

Of course it is unlikely that those with power and influence, either on the environmental or economic front, will willingly cede the battle to a process of adaptive discovery that could very well undermine their power and influence. What if economic vitality turns out to be much more of a concern in the short to medium term? What if it turns out that environmental integrity is the concern?

Many U.S. cities today possess a legacy of facilities imperfectly suited to their transportation requirements. This legacy reflects the assumptions of prior generations of transportation planners about what long-term requirements would be. Indeed, those facilities provide a remarkable level of mobility—to those with the wherewithal to live in places served well and to own and operate the necessary vehicles.

The challenge is how to move forward. Environmental activists have effectively blocked most efforts to increase capacity. The slack capacity built in prior generations is rapidly running out. Advanced communication technology can help. But a showdown over expanding the urban highway network looms on the horizon.

One alternative is to let the showdown occur. This is messy. But in a democratic, market-based society like that of the U.S., the eventual result will probably be approximately in line with public preferences. It also promises some good theater.

But this alternative also harbors some risks. Transportation is a critical economic service that affects the distribution of goods and accessibility to jobs, housing, and services. A showdown would almost certainly harm transportation in the short term. For a century, the U.S. has been increasing its utilization of the speed and autonomy afforded by road transport. A showdown is likely to interrupt that trend. Indeed, that is exactly the objective of the reformers. The economic dislocations of this interruption are difficult to predict. However, the oil supply restrictions of the 1970s and the reduced capacity of the transportation system following the terrorist attacks of September 11, 2001, are sobering reminders of the potential sharpness of their economic impacts.

A more troubling risk is that the resolution of the showdown will reflect the interests, assumptions, and beliefs of those with power and influence today, which will almost certainly fail to accord with uncertain future technological, economic, social, and environmental conditions.

Adaptive Discovery

A better choice is adaptive discovery, whereby resolution of transportation and environmental problems can reflect the values and priorities of those closest to the situation. After all, most transportation and environmental problems are local and regional.

Facilities such as roads, rail lines, and bridges are expensive and long lasting. How much room is there for flexibility and adaptation in their planning and design? The Austrian concept of capital goods as "fossils" of prior capital investment plans (see the previous chapter) suggests that the key strategy is to redeploy capital goods when expected conditions fail to materialize as quickly as new information becomes available. This is a feedback or learning process, from expectation to realization to revised expectations to new realizations—in short, a discovery process.

How does such redeployment work in the case of urban highways and transit? End users adapt quickly, discovering any facilities that can help them speed their travel, regardless of the intentions of planners.

Traffic operations officials typically do their best to deal with what happens. They are concerned about identifying and clearing accidents, dealing with facility maintenance activities that require lane closures, and so on. They are on the front line, and when they err they usually hear about it quickly.

What about planners? Current planning practice tends to emphasize conforming to prior plans, which usually include future resource commitments to promised projects many years into the future. Once a project is in a long-range plan or on a list of funded projects, it may be extremely difficult to dislodge that project in light of new information, that is, to adapt the plan to emerging conditions. Project advocates are usually entrenched, and all parties to the plan have an interest in preserving its inviolability lest one of their projects be the one considered for modification or deletion. Horse trading and back scratching have often played a part in nominating projects to the list, and deleting a project would violate the deals and commitments to other projects on the list.

This is equally true for project advocates and project opponents. Anything that disrupts these compromises is viewed as unwelcome. A case in point: following the 2001 terrorist attacks, New York's governor requested a four-year waiver of Clean Air Act restrictions on project construction. Environmental interest groups strongly opposed the waiver. The governor claimed that many planned projects either no longer made sense or had lessened in priority; in response, environmental interest groups claimed that any waiver would allow the crisis to be used as a subterfuge for advancing environmentally unsound projects.[8]

As a result, the planning process can end up focusing more on preventing adaptation and response to new information than on making wise decisions based on the best information available when a decision can no longer be deferred. Of course, advocates of technically objective transportation planning have long decried the influence of politics in the selection of projects. This book's call for reform is not so naïve as to think that politics will go away, nor, indeed, to advocate that it should. In a democracy elected officials, not career civil servants or professional planners, are the ones charged with the difficult job of representing the people's value tradeoffs.

But there is value in a more flexible approach that strives to keep options open, defers committing to a single option until it is imperative to act, and then makes the technical and political judgments based on the best information available. The current approach, on the other hand, seems designed to force early commitment to decisions long into the future with minimal capability to change course in light of future, better information.

The author has already heard howls of outrage from some in the transportation community that the last thing society needs is a mechanism to allow politicians to avoid making tough choices.

---

[8] Richard Pérez-Peña, "Pataki Seeks U.S. Waiver on Air Quality," *New York Times*, January 2, 2002.

Indeed, many argue that what is needed is a more rigorous method of committing to long-term facility decisions and preventing any way to "weasel out" in the face of controversy about project impacts on parks, neighborhoods, or other concerns. The argument, in short, is that the benefits of large projects are far in the future, and that it is necessary to withstand short-term pain to realize long-term benefits. Politicians, the argument goes, elected in the here and now, have no long-term view, and hence are inappropriate agents of the long-term interests of society.

This book challenges this line of reasoning on two counts. First, there is no reason project benefits have to be far in the future. While some projects take decades to complete, much of the delay in project development is a result of procedural impediments put in place precisely because projects were unsuitably scaled and designed. The initial construction of the urban interstates was completed quickly, and their benefits redounded immediately to the residents of many cities. But it also soon became apparent that their design was so damaging to the surrounding communities that elected officials intervened and interjected other considerations into the process. Indeed, many cities around America today are spending billions of dollars tearing down and reconstructing facilities of that era, and many other cities wish they had the congressional clout to get funds earmarked for their projects. The error was not in the shortsightedness of the elected officials of that day; indeed, they intervened to protect their cities from misguided and inept designs.

A second flaw in judging elected officials incapable of representing the long-term interests of society is the misguided assumption that the projections of net benefits produced through the planning process are a suitable proxy for long-term societal interests. As we have argued in previous chapters, it just ain't so.

In fact, an emphasis on short- and medium-term localized impacts is not as radical as it may sound. As urban transportation planning has moved beyond sixties- and seventies-style large-scale regional planning, it has begun to focus more on site-specific and corridor-specific problems. In the 1980s, federal initiatives sought to lessen the burden of federal regulation, give more flexibility to local areas, and provide guidance on good practice rather than specific prescriptive procedures.[9] The emphasis on good practice over specific prescription continued and is reflected in the urban planning regulations promulgated under ISTEA.[10]

Attention to the short and medium term is also apparent in the growing prominence of operations and management (O&M) in transportation agencies. Increasing attention to O&M is the result of a confluence of forces. One is a declining volume of new design and construction projects with the completion of the Interstate and an increasing antipathy to new construction among community and environmental interest groups. This has accompanied a commensurate

---

[9] Edward Weiner, *Urban Transportation Planning in the U.S.: An Historical Overview*, DOT-T-93-02, rev. ed. (Washington, D.C.: U.S. Department of Transportation, November 1992), 206-34.
[10] U.S. Department of Transportation, "Statewide Planning; Metropolitan Planning," final rule, *Federal Register* 58, no. 207 (October 28, 1993): 58040-79. See also, Cambridge Systematics, Inc., *Multimodal Transportation Development of a Performance-Based Planning Process*, Research Results Digest (National Cooperative Highway Research Program), no. 226 (Washington, D.C.: National Research Council, Transportation Research Board, July 1998).

decline in design and construction as the most prestigious areas of transportation agency business. Continued increases in road traffic without corresponding increases in road space have also heightened concern about operating facilities efficiently. The advent of intelligent transportation systems (ITS) since the late 1980s has also brought attention to facility and system O&M in areas such as traffic signal control systems and enhanced incident detection and incident response.

Emerging interest in asset management has also focused attention on O&M. Asset management is an approach that focuses on managing facilities to minimize costs and maximize benefits. It relies on improved data on facility condition and user requirements. Capital planning under asset management grows out of operations and management, as opposed to separate master planning activities.[11]

Heightened attention to O&M is reflected in the Federal Highway Administration's creation of an Operations Core Business Unit in 1998. The 2001 National Dialog on Transportation Operations and the associated National Summit on Transportation Operations also sought to focus the attention of the transportation community on these issues.[12]

A shift towards more short- to medium-term planning horizons in clean air policy, which seeks to reinforce the role of long-term regional-scale planning models. These models are a key input into the determination of a region's conformity with the Clean Air Act. Environmental interest groups have supported national anti-sprawl initiatives focusing on long-term regional growth,[13] and numerous "smart growth" and "sustainable development" initiatives have been launched.

To some extent, environmentalists' reliance on long-term models is a reaction to the past dominance of such models in setting the transportation agenda. Those models should not have ignored or subordinated clean air and other environmental impacts. But concern about the environment is no justification for an unsuitable long-term orientation for transportation decision making.

Instead, now that the playing field is more level, all parties should work towards collapsing long-term planning horizons and developing shorter, more manageable timelines. Recent efforts in environmental streamlining are a promising beginning.[14] A process of adaptive discovery holds the greatest promise for identifying and realizing the proper combination of economic vitality and environmental integrity, and the proper combination of control and flexibility.

* * *

This chapter argues that a new approach to transportation planning is urgently needed. The current approach focuses excessively on ensuring adherence to prior plans and preventing adapta-

---
[11] Odd J. Stalebrink and Jonathan L. Gifford, "Actors and Directions in U.S. Transportation Asset Management," paper presented at the Transportation Research Board 81st Annual Meeting, January 13-17, 2002, Washington, D.C.
[12] See, for example, Ken Orski, "Developing an 'Operations Vision': The Dialog Continues," *Innovation Briefs* (Washington, D.C.: Urban Mobility Corporation) 12, no. 2 (March/April 2001).
[13] Justin Blum, "Anti-Development Campaigns Take Root," *Washington Post*, July 29, 1998, B1ff.
[14] U.S. Department of Transportation, *Highway and Transit Environmental Streamlining Progress Summary*, Report to Congress (February 2002).

tion to new conditions, situations, and information. The next chapter turns to the difficult matter of specifying a new approach.

CHAPTER APPENDIX: BIBLIOGRAPHY ON CONCEPT OF FLEXIBILITY

Barns, Ian. 1991. Post-Fordist People? Cultural Meanings of New Technoeconomic Systems. *Futures*, November: 895-914.

Benjaafar, Saifallah, Thomas L. Morin, and Joseph J. Talvage. 1995. The Strategic Value of Flexibility in Sequential Decision Making. *European Journal of Operations Research*, 82: 438-57.

Boer, Harry, and Koos Krabbendam. 1992. Organizing for Manufacturing Innovation: The Case of Flexible Manufacturing Systems. *International Journal of Operations & Production Management*, 12, no. 7/8: 41-56.

Bunasekaran, A., T. Martikainen, and P. Yli-Olli. 1993. Flexible Manufacturing Systems: An Investigation for Research and Applications. *European Journal of Operations Research*, 66: 1-26.

Crowe, Thomas J. 1992. Integration Is Not Synonymous with Flexibility. *International Journal of Operations and Production Management*, 12, no. 7/8: 26-33.

Eccles, Henry E. 1965. *Military Concepts and Philosophy*. New Brunswick, N.J.: Rutgers University Press.

Gertler, Meric S. 1988. The Limits to Flexibility: Comments on the Post-Fordist Vision of Production and Its Geography. *Transactions of the Institute of British Geographers*, n.s., 13: 419-32.

Gerwin, Donald. 1987. An Agenda for Research on the Flexibility of Manufacturing Processes. *International Journal of Operations and Production Management*, 7, no. 1: 38-49.

Haughton, G., and J. Browett. 1995. Flexible Theory and Flexible Regulation: Collaboration and Competition in the McLaren Vale Wine Industry in South Australia. *Environment and Planning A*, 27: 41-61.

James, Jeffrey, and Ajit Bhalla. 1993. Flexible Specialization, New Technologies and Future Industrialization in Developing Countries. *Futures*, July/August: 713-32.

Kapur, S. 1992. Of Flexibility and Information. University of Cambridge Economic Theory Discussion Paper no. 77, Cambridge, May.

Lim, S. H. 1987. Flexible Manufacturing Systems and Manufacturing Flexibility in the United Kingdom. *International Journal of Operations and Production Management*, 7, no. 6: 44-54.

Lindberg, Per. 1992. Management of Uncertainty in AMT Implementation: The Case of FMS. *International Journal of Operations & Production Management*, 12, no. 7/8: 57-75.

Malecki, Edward J. 1991. The Location of Economic Activities: Flexibility and Agglomeration. In *Technology and Economic Development: The Dynamics of Local, Regional and National Change*, chap. 6. New York: John Wiley & Sons.

———. 1995. Guest editorial. *Environment and Planning A*, 27: 11-14.

Mandelbaum, Marvin, and John Buzacott. 1990. Flexibility and Decision Making. *European Journal of Operations Research*, 44: 17-27.

Merkhofer, M.W. 1977. The Value of Information Given Decision Flexibility. *Management Science*, 23: 716-28.

Murdoch, J. 1995. Actor-Networks and the Evolution of Economic Forms: Combining Description and Explanation in Theories of Regulation, Flexible Specialization, and Networks. *Environment and Planning A*, 27: 731-57.

Newman, W. Rocky, Mark Hana, and Mary Jo Maffei. 1993. Dealing with the Uncertainties of Manufacturing: Flexibility, Buffers and Integration. *International Journal of Operations and Production Management*, 13, no. 1: 19-34.

Pye, Roger. 1978. A Formal, Decision-Theoretic Approach to Flexibility and Robustness. *Journal of the Operational Research Society*, 29, no. 3: 215-27.

Quinn, James Brian. 1978. Strategic Change: "Logical Incrementalism." *Sloan Management Review*, Fall: 7-21.

———. 1989. Strategic Change: "Logical Incrementalism." Reprint, with comment. *Sloan Management Review*, Summer: 45-60.

Skorstad, Egil. 1991. Mass Production, Flexible Specialization and Just-in-Time: Future Development Trends of Industrial Production and Consequences on Conditions of Work. *Futures*, December: 1075-84.

U.S. Federal Highway Administration. 1997. *Flexibility in Highway Design*. Washington, D.C.: U.S. Dept. of Transportation, Federal Highway Administration.

9

TRANSPORTATION PLANNING: A FLEXIBLE APPROACH FOR THE TWENTY-FIRST CENTURY

The challenge of proposing a new approach to urban transportation planning is that there are no easy answers. Any effort to reinvent, or simply improve, current practice confronts the stark reality that planning and operating agencies rely on current practice—with all of its weaknesses—to get things done on a day-to-day basis. "Don't shoot the horse I'm riding on," implored one metropolitan planning official at a recent conference on improving the planning process. While she recognized the need for improvements, her agency also needed a canon of accepted "best practice" to defend projects challenged in court by environmental interest groups. In many ways, however, transportation planning has become more a highly stylized ritual of procedures and challenges than an objective, scientific problem-solving activity.

Debate about urban transportation planning often focuses on analytical models and public participation. Despite the widespread use of traditional models, the profession generally recognizes their flaws and unsuitability for assessing many of the policy problems to which they are applied. To address this concern, a large federally funded transportation research effort is underway to improve the modeling system.

Concern about the integrity of the public participation process is also widespread, although perspectives on remedies differ sharply. Both transportation reformers and infrastructure advocates view each other as wielding too much influence.

What is really going on in many transportation debates, however, is conflict over values, which cannot be resolved through more sophisticated analytical models or better participation guidelines. When goals conflict, arguing about analytical models is not constructive. Neither is establishing more rigorous procedures controlling who may comment on the planning process and when. Better models and better technical procedures alone cannot resolve conflicts over the "right" future for urban transportation.

Efforts to improve both models and participation fail to examine the questions being asked in the planning process. They emphasize how given questions can best be answered. But the questions themselves fail to articulate for deliberation and debate the core value conflicts that are at the center of most debates over urban transportation planning.

The objective of a new, flexible approach to transportation planning is a planning and decision process that recognizes these core value debates and, where possible, helps resolve them. Debate should not concentrate on the procedures for developing competing long-term forecasts and the participants who decide among them. Instead, it should recognize the inherent unpredictability of the long-term future and focus on near-term problems and opportunities.

Transportation planning must be flexible because it must recognize that no single objective truth exists to guide planners and decision makers unambiguously to what is right. Rather, communities, households, and firms must themselves choose between competing values through their behavior in markets and participation in political processes.

In many areas such an approach will favor mobility. In others it may favor community and environmental preservation. The appeal of this approach is that it places decision making where it needs to be: at the level of the household, firm, and community. It also promises to end at least some of the conflicts so society can focus on other important determinants of quality of life and economic prosperity.

This is not to say that refining and improving planning and decision models should cease. Far from it! Many able researchers are focusing on that problem.

Better models cannot, however, resolve value differences. When there is consensus on goals, models can help refine and support decision making. But when goals differ, debates over analytical models can distract from the decisions and compromises necessary to move forward.

Important as they are, debates over values are often indirect and elliptical. Engineers are often uncomfortable with them. Value differences do not lend themselves to objective, dispassionate debate and resolution. But the failure to address value conflicts squarely leads to stalemate: It hamstrings further development of infrastructure to serve the economy and broader society and impedes changes in public policy and public behavior that could allow the more efficient use of existing infrastructure and obviate the need for more of it.

STABILITY AND AGILITY

Social process, public policy, and—of particular interest here—transportation policy must strike a balance between stability and agility. Stability is an anchor, embodied in heritage, culture, customs, habits, routines, traditions, property rights, entitlements, artifacts, and institutions. Agility is opportunistic, responsive, flexible, and adaptable. The companion of stability is predictability. The companion of agility is uncertainty.

Stability and agility coexist in dynamic tension. Agility can carry society forward. Stability can prevent it from sliding backwards. Planning cannot, and should not, resolve this conflict in any final way. Striking the right balance between them is a social process that necessarily involves conflict, rivalry, and learning.

The current emphasis on planning procedures and participation has to a great extent immobilized planning and decision making in many regions. Meanwhile a showdown is looming between the advocates of environmental integrity and economic vitality as existing facilities, especially highways, approach capacity.

A new, flexible transportation planning process can help foster coexistence between competing views and strike the right balance between them. Some rivalry and conflict is inevitable, and is indeed healthy, as it can foster and facilitate this process, without necessarily biasing it.

Progressive ideology has predisposed decision-making towards predictability, which has led to bias towards particular outcomes. The appropriate balance between stability and agility can only be worked out by users and other stakeholders, not through engineering and modeling.

REINVENTING TRANSPORTATION PLANNING

A new transportation planning process that embraces flexibility is at one level a simple elaboration of the common rational or systems model. It would support decision making with several functions:

Intelligence: collecting raw current and historical data, developing methods of displaying and summarizing them, developing methods of analyzing and projecting them.

Decision support: facilitating group processes, convening meetings, hosting decision making processes, analyzing alternatives. (When there is consensus on goals, e.g., accommodating traffic growth, preserving open space, then the decision support function could systematically translate those goals into concrete plans. When there is no consensus, it can generate alternatives and seek to identify common ground. The intent is to be policy neutral.)

Design and implementation: taking project plans and moving them towards implementation. May involve right-of-way reservation.

Monitoring and review: maintaining flexibility, conducting flexibility reviews.

A Commitment to Honesty

A central tenet of this new planning is honesty—honesty about what is known and knowable, and honesty about what is subjective and judgmental. Alas, this differs from common practice, which tends to focus on advocacy of competing versions of objective truth. The tools for formulating and defending those competing versions of truth are models, experts, forecasts, and projections.

In transportation planning a lively debate has begun about honesty in the modeling and decision-making process, focusing on the truth and accuracy of models for assessing and evaluating competing policy proposals and retrospective analyses of investments and decisions. Consultants state privately that clients threaten that the "wrong" results of an analysis will result in unfavorable consideration for future contracts or the nonrenewal of existing contracts. Financially stronger firms can resist such influence, but financially weaker ones may not be able to do so. Some have suggested instituting peer review of project assessments to protect consultants from such undue influence.[1]

A key battleground has been the debate over forecast and actual cost and ridership of the new urban rail transit systems initiated in the 1960s and 1970s. A 1989 federal study concluded that ridership forecasts were too high and cost figures too low, ultimately leading to sharply higher per passenger costs. The report attributed this systematic error to bias in the cost and passenger forecasting process.[2] This report unleashed a furious debate within the field about the ethics of the forecasting process.[3]

However, this debate has largely focused on the technical veracity of forecasts—that is, on refining and improving the forecast instruments themselves, increasing the influence of expertise in making decisions, and reducing the role of politics. It has not raised the more fundamental question about the role of forecasts in the decision process—that is, are forecasts an appropriate instrument for making decisions given the considerable uncertainties and contingencies facing planners.

Answering the question of whether such transit systems should have been built, and whether more should be built in the future, requires much more than a technical answer. It involves a debate about values—values concerning economic efficiency, automobile dependency, settlement patterns, preservation of open space, and the viability of pre-World War II central business districts.

The way to resolve such debates is not to argue about how to ensure technically valid forecasts. It is to be honest about the uncertainties and contingencies associated with the decision and to foster a debate and resolution about which values, in a particular instance, are to have primary and which secondary status.

Thus a key feature of a flexible approach to transportation planning is a new role for experts in the decision-making process. Expertise has tended to focus on those aspects of a problem that yield to expertise.

---

[1] Martin Wachs, "Ethics and Advocacy in Forecasting for Public Policy," *Business and Professional Ethics Journal* 9 (1990): 141-57; Martin Wachs, "When Planners Lie with Numbers," *Journal of the American Planning Association* 55, no. 4 (1989): 476-79; John F. Kain, "Deception in Dallas: Strategic Misrepresentation in Rail Transit Promotion and Evaluation," *Journal of the American Planning Association* 55, no. 4 (1989): 476-79.
[2] U.S. Department of Transportation, Urban Mass Transportation Administration, *Urban Rail Transit Projects: Forecast versus Actual Ridership and Costs* (Washington, D.C., 1989).
[3] Special issue on transportation forecasting, *TR News* 156 (September-October 1991).

Transportation experts must recognize that decisions about the future of a region's transportation system do not simply involve answering technical questions, and that political processes are a legitimate mechanism for answering them. While technical analysis is an important element, so too is subjective judgment, which experts alone cannot resolve. The profession has a responsibility to stop pretending that it knows what it does not know. That is the first step.

Speedy decisions in the absence of consensus on goals are not easy under the American system.[4] With consensus, in an emergency it is possible for government to act quickly and use its coercive power to great effect. The response to the terrorist attacks of September 2001 is a case in point. Invasive and time-consuming security policies that travelers and the airline industry had resisted for years were suddenly acceptable. In the absence of consensus, however, government takes decisions slowly, incrementally, and tentatively.

Flexible planning will not create the illusion of consensus where none exists, nor speed up decision making in the absence of consensus. Rather, it will focus attention on differences in core values, facilitate their resolution where possible, and acknowledge differences that are irreconcilable at present. Unlike current practice, flexible planning does not involve arguing about a chimerical set of definitive models; instead it focuses on value differences where they exist.

There are two aspects to the focus on honesty. One is false certainty, the recognition of what is not known—the predictive modeling fallacy. Equally important is honesty about what is known, what is known to work, and not work, what is technically feasible. While every point of view deserves consideration in a public proceeding, professionals have a responsibility to point out ambiguity, falsehood, and unclear thinking, and to distinguish between what is and is not factually supportable.

One argument in the debate over transit ridership and costs, for example, was that transit ridership forecasts were based on assumptions about zoning and high-density development around transit stations that were based on predictions of high-density development around transit stations that failed to materialize, due to public opposition or unfavorable market conditions. That is true in some cases. Thus investments that rely heavily on future zoning decisions ought to be made with the awareness that zoning and land use decisions are not entirely predictable and that communities and their public officials may choose to prevent the high-density developments necessary to realize the transit forecasts.

To a large extent the last three decades' attempts to attenuate the American public's desire for private travel have not worked. While use of alternative travel modes have increased here and there, emphasis on high-occupancy vehicle lanes and expanded transit service has almost universally failed to attract a growing market share. Advocates attribute that failure variously to the lack of fully connected HOV (high occupancy vehicle) networks and continued subsidies of the automobile. But a more likely explanation is that consumers are not willing to endure the inconvenience of HOV or transit unless the alternative of driving alone reaches levels of discomfort and inconvenience that are not politically viable. Moreover, regulatory constraints im-

---

[4] I am indebted to my colleague Brien Benson for articulating this point.

posed on transit and shared-ride services in most American cities constrain the ability of current and potential transit suppliers to offer services that are economically viable.[5] These explanations suggest emphasis on alternatives with the potential to make a significant dent in total and peak hour demand, like four-day workweeks, telecommuting, and regulatory reform of transit, rather than continued emphasis on proven losers.

Truth, then, is a first principle. Acknowledge uncertainty, ambiguity, and contingency on future decisions and actions. Reject assertions that cannot be supported with facts. That means dismantling the web of models and invalidated assumptions that support so much decision making today. This will be painful. It will take years. But it is an essential first step.

Intelligence: A Source of Factual Information

An essential function for flexible planning is reconnaissance: collecting and disseminating information about current conditions in transportation and related areas. This could include a broad range of activities (see Table 9-1). At the low end, agencies could simply make available data they already collect as a byproduct of operations and monitoring. This would include traffic counts on streets and highways within a jurisdiction, surveys of pavement conditions, ridership counts, budget information, and spending records. At the high end, it could include tools for summarizing and displaying data, drawing from geographical information systems (GIS) and other analytical and graphical tools.

The type of data made available could be quite diverse (see Table 9-1). Traffic counts on transit and highways are routinely collected, at least episodically, by most operating agencies. An intelligence system could make such counts available. It would be important to specify details of the collection time and date. Though data on pedestrian and bicycle traffic is typically slim, the recent heightened interest in fostering the use of nonmotorized transport suggests the need for better data in this area.

Speed is one of the most common and easily understandable measures. A rich measure would include average speed and speed variance, both within a trip and from trip to trip. The trip-to-trip speed variance would also serve as an indicator of service reliability.

Safety is another important indicator that includes data on fatalities, injuries, accidents, and property damage, and might also include demographic characteristics of those involved in incidents, such as age, race, and gender. Care would be necessary to ensure the privacy of individual data.

---

[5] Clifford Winston and Chad Shirley, *Alternate Route: Toward Efficient Urban Transportation* (Washington, D.C.: Brookings Institution Press, 1998); Daniel B. Klein, Adrian Moore, and Binyam Reja, *Curb Rights: A Foundation for Free Enterprise in Urban Transit* (Washington, D.C.: Brookings Institution Press, 1997).

Energy indicators could measure the consumption of energy (btu/trip, btu/mile), as well as the fuel used (e.g., petroleum, solar). Environmental impact could include noise, air quality, water quality (drainage, erosion, runoff pollution), and wildlife (animals killed).

Table 9-1
Example Factual Data Categories

Traffic counts
Speed
Safety
Energy consumption
Congestion
Availability
Pavement condition
Budget and actual spending

Congestion is another dimension of transportation system performance, which received considerable attention in the 1990s. The Intermodal Surface Transportation Efficiency Act of 1991 (ISTEA) mandated that states establish a number of management systems, including a congestion management system (CMS). While Congress subsequently repealed the mandate for the management systems in 1995, it sparked considerable interest in congestion measurement while it was in force.[6]

A working definition of congestion is "travel time or delay in excess of that normally incurred under light or free-flow travel conditions." Unacceptable congestion is travel time or delay in excess of an agree-upon norm, which may vary by type of transportation facility, geographic location, and time of day.[7] Clearly, such a definition requires objective information on travel time and delay, as well as information on "agreed-upon norms." The latter may go beyond the capability of an intelligence function, but an intelligence function would certainly provide objective information about travel time and about delay information such as queue lengths.

Availability measures the amount of prearrangement required for particular services or facilities. Services that operate on a schedule or require advance reservation are less available, whereas those that can be used on demand are more available. Availability measures would indicate the amount of advance notice required the waiting times associated with access to a particular service or facility, and how they vary by time of day, etc.

While the technical resources needed to make such information available may be modest, the human resources to put data in an electronically accessible format may be considerable. Some data are generated electronically and could simply be made available. Other are generated and

---

[6] "Intermodal Surface Transportation Efficiency Act of 1991," P.L. 102-240, *Stat.* 105 (18 December 1991): 1958; "National Highway System Designation Act of 1995," P.L. 104-59, *Stat.* 109 (28 November 1995): §205.
[7] Tim Lomax et al., *Quantifying Congestion*, Final Report, Prepared for National Cooperative Highway Research Program, Project 7-13, Transportation Research Board, National Research Council (College Station, Tex.: Texas Transportation Institute, September 1995), p. 15.

182 FLEXIBLE URBAN TRANSPORTATION

recorded mechanically or manually, and their conversion into electronic format would require an effort. The recent addition of the archive data user service to the ITS architecture, for example, reflects an awareness of the potential for ITS data to support an intelligence function.

An agency wishing to initiate an intelligence function under a flexible planning regime would need to prioritize its efforts, cataloging what data would be useful to the public, estimating the cost of codifying and making such data available, and maintaining the data stream once it is made available. The Internet is an obvious mechanism for making such data available, with web sites developed either by some more advanced agencies and their consultants, or under contract from the federal government. An agency might convene an advisory committee to help prioritize data availability efforts.

Making raw data available to the public without providing information about what the data's significance, biases, and inaccuracies may be a cause for concern. However, the view of an EPA official on this matter is instructive: "'[T]he availability of data to the public [is] one step in educating the public a little more, and closing the gap between how the experts look at the problem and how the public views it.'" The official also indicated that he thought the public's analysis of such data, environmental risk data in this case, was more sophisticated than the agency's because of the public's ability to address the full range of implications without the constraints imposed by agency rules, procedures, and conflicting constituencies.[8]

An additional concern is the security and safety of transportation facilities and services. Data on the location and operational characteristics of transportation systems could be misused.

Making data available to the parties interested in transportation policy and planning could also help inform the public and close the gap between expert views of problems and the public's views. Such data sets could build on or be derived from some of the available on-line sources of traffic and travel information already being offered. For example, World Wide Web sites currently offer real-time traffic congestion information in many cities. The rapidly dropping cost of storing the complete history of such data series makes data storage more feasible. If complete archiving is not possible, summary statistics could be substituted.

Decision Support

The intelligence function of a flexible planning program focuses on making raw data available to interested parties, possibly along with tools for summarizing, describing and displaying such data. Intelligence stops short of any analysis or modeling of raw data, however. These would be the province of a decision support function.

---

[8] Whereas EPA's analysis of the data had to follow tightly defined procedures that resulted from often contentious debate about the scientific validity of procedures, the public was free to view the data from a full range of perspectives. While public analyses of the data might also be "unscientific," the contentiousness of the debates within EPA and its stakeholders suggest that a standard of scientific objectivity might not always hold sway in the debate over their procedures either. EPA official quoted in Malcolm K. Sparrow, *The State/EPA Data Management Program*, Parts A-D (Cambridge: Harvard University, Kennedy School of Government, 1990), part B, p. 13.

Transportation Planning 183

The services available through a decision support function could cover a broad range, from tools for summarizing, describing data, and extrapolating data to a full range of services, such as modeling tools like traditional four-step transportation planning models, economic models, and environmental models.

The decision support function might be further developed where there is consensus on goals. In such cases, a decision support function could provide detailed tools for simulation and evaluation, similar to those currently used and envisioned in much of the mainstream transportation planning literature. Such tools, however, may not be very useful if agreement on goals is absent.

Figure 9-1
The Competing Roles of Planning

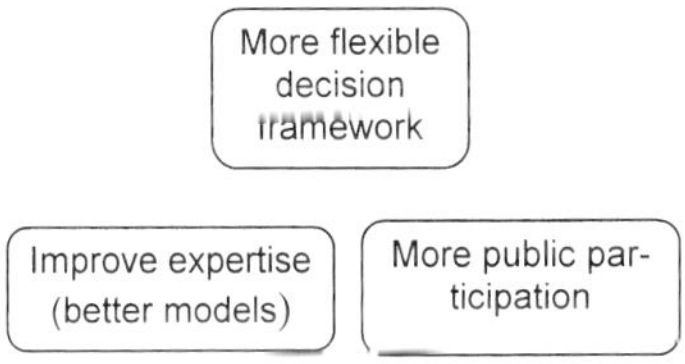

```
                ┌──────────────┐
                │ More flexible│
                │   decision   │
                │   framework  │
                └──────────────┘

   ┌──────────────────┐   ┌──────────────────┐
   │ Improve expertise│   │ More public par- │
   │  (better models) │   │    ticipation    │
   └──────────────────┘   └──────────────────┘
```

Thus, the role of planning will not be static or uniform. When there is consensus on goals, expertise could dominate. When a community has little consensus on goals, more public participation would be indicated (see Figure 9-1). When the behavior of the underlying system is well understood, then expertise can play a big role. When the underlying system is not easily understandable, then ritual—perhaps competing ritual—needs to play a role. The process may not be particularly rational or objective. Multiple interpretations may prevail. The framework in Figure 9-2 (also discussed in the last chapter) illustrates again that agreement about both goals and the nature of the underlying system are required for analytically oriented techniques to be useful.

Transportation planners must recognize when the degree of consensus is waxing and waning and act accordingly. They should not assume consensus on goals. They should also recognize that better models, such as the Federal Highway Administration's Travel Model Improvement Program (TMIP), while they may have value at some level, will not help achieve consensus on values—which are often at the heart of transportation policy debates. Conflicts about urban form and the role of transit will not be resolved through microsimulation.

One of the limitations of a decision support function is that such models often require a training and technical sophistication to operate and interpret. This might not be available in jurisdictions where the planning function is only minimally supported. There are a variety of ways to address this problem. Wealthier jurisdictions might offer advanced technical support, possibly a portion of full position, or even multiple positions, depending on the level of public demand. Support technicians could provide a range of support, from helping find relevant data locally

184 FLEXIBLE URBAN TRANSPORTATION

and in other communities for comparison to running various models to compare and contrast alternatives. Full-time staff and permanent space would rarely be appropriate, since public interest in transportation issues tends to wax and wane. So support personnel would likely have other responsibilities.

Figure 9-2
Framework for Goals and Means

Agreement about goals

	1. Analytic techniques	3. Negotiated
	2. Pragmatic	4. Adaptive discovery

Agreement about means to achieve goals

Source: Adapted from Russell Stout, "Management, Control and Decision," in *Management or Control* (Bloomington: Indiana University Press, 1980), 101.

Another model to consider would be a cooperative effort with a local university. A state might contract with one or more of its state or private universities to provide a support and outreach function. Yet another option would be to contract out such a service to a consultant or nonprofit group.

Design and Implementation

Once decision makers arrive at a decision, perhaps with the aid of the decision support function, emphasis must shift to design and implementation. A flexible approach is necessary here as well. No hermetic seal exists between decision making and implementation, of course. Discoveries and learning in design and implementation may lead to reconsideration of prior decisions. Indeed, any healthy decision-making process allows and encourages such feedback.

The typical transportation planning and programming process is not well suited to using discoveries in design and implementation to modify and condition future decisions. It can be difficult to modify or abandon a project if detailed design yields unpleasant surprises. The complex multi-agency project review and approval process strongly discourages any reconsideration of prior decisions about project scale and scope or reprogramming of funds.[9]

The specific features of a community's support of design and implementation activities will vary widely, depending on the nature of the community and the project under consideration. Advisory committees and design review teams will be used on some projects as integral parts of the design and implementation process. Other projects will depend primarily or exclusively on staff and contractors. Two aspects of design and implementation are worthy of separate consideration: design and procurement, and outreach and community involvement.

Design and Procurement

Flexibility in design and procurement has received increasing attention in recent years. Part of the reason is rapid innovation and infusion of new technologies into road transport, including Intelligent Transportation Systems (ITS) and improved pavement technologies. Designs that fail to accommodate innovations that may become feasible, either in the course of the project implementation or after it opens for service, can increase project costs.

Incorporating such flexibility into the traditional "stovepipe" project development process used in many agencies has been difficult, however. Under the stovepipe approach, planning, design, construction, and operation are separate organizational activities. Design and construction are often contracted out, typically to unrelated contractors. Hence, the design contractor produces a design, which the sponsoring agency then uses as the basis for awarding contracts for construction. Changes in design typically require change orders, which are outside of the competitive process and therefore often expensive.

The construction of the Central Artery/Tunnel project in Boston is an example. This $10+ billion project was originally designed required that each tollbooth accommodate attended collection, unattended coin collection, and automatic vehicle identification (AVI). Implementations of AVI equipment around the United States proliferated rapidly in the early 1990s, and the desirability of AVI-only lanes became apparent. Modifying the design to incorporate AVI-only lanes required change orders for five design packages, one of which was already in construction. Processing change orders was cumbersome and time consuming.[10] This example demonstrates an axiom of transaction cost economics: unless an agreement is properly specified to avoid holdups, a party to a contract (such as a design or construction contractor), not a public agency and its constituents, can appropriate almost all of the benefits of such opportunities for improvement.[11]

---

[9] Kingsley E. Haynes et al., "Planning for Capacity Expansion: Stochastic Process and Game-Theoretic Approaches," *Socio Economic Planning Science* 18, no. 3 (1984): 195-205.
[10] Serge Luchian, Daniel Krechmer, and Paul Muzzey, "Case Study of Electronic Toll Collection in the Central Artery/Tunnel Project—Boston," in *Infrastructure Planning and Management*, ed. Jonathan L. Gifford, D. Uzarski, and S. McNeil (New York: American Society of Civil Engineers, 1993), 561-72.
[11] Paul H. Rubin, *Managing Business Transactions: Controlling the Cost of Coordinating, Communicating, and Decision Making* (New York: Free Press, 1990).

Many agencies are exploring alternative service delivery approaches, such as design-build-operate, whereby an agency awards a single contract to an entity that designs, constructs and operates a facility for a period of time. In some cases, agencies specify a particular level of performance, say a given level of traffic to be served, and leave the details of how to provide that to a contractor. Such an approach was used, for example, in letting a contract for an 8-mile (12.9-km) bridge between New Brunswick, on the Canadian mainland, and Prince Edward Island. Under the contract, the government specified a fixed, inflation-adjusted payment of C$ 41.9 million in 1992 dollars to a franchisee for thirty-five years. Performance requirements included all-weather operation for 12 months a year. In another case, electronic toll collection for Highway 407 in Toronto, Canada, was procured on a performance basis (99.99997 percent accuracy), and the technological details were left to the contractor.[12]

The American Association of State Highway and Transportation Officials (AASHTO), the Federal Highway Administration (FHWA), and the Transportation Research Board (TRB) have all initiated projects to review and identify better technology transfer, better use of research findings, and more flexible procurement and operation options.[13] State departments of transportation in the U.S. are beginning to explore such alternative program delivery approaches.[14]

Such approaches can allow for more risk taking and more rapid adoption of innovations. Part of a flexible approach is recognizing that traditional procurement and project development approaches may be inappropriate.

Outreach and Community Involvement

Another important element of flexible design and implementation is the role of outreach and participation. Large-scale projects usually have large-scale impacts on communities, households, businesses, the environment, historic buildings, and land uses. The nature of those impacts depends heavily on how a project is designed and implemented. The old practice of designing a project on the basis of engineering criteria, and then condemning the property needed to build it has long since given way to a more sensitive, interactive approach. Communities are informed of project progress, and public input is often solicited to help refine project design and mitigate project impacts, both in the construction and operation phase.

The construction of Interstate 66 inside the Washington Beltway is an example. This project, which had been stalled for decades in the environmental review process, was finally completed

---

[12] John B. Miller, *Principles of Public and Private Infrastructure Delivery* (Boston: Kluwer, 2000), 209-22.
[13] Transportation Research Board, *Innovative Contracting Practices*, Transportation Research Circular, No. 386 (Washington, D.C.: Transportation Research Board, National Research Council, 1991); American Association of State Highway and Transportation Officials, *European Asphalt Study Tour* (Washington, D.C., 1990); U.S. Federal Highway Administration, *Report on the 1992 U.S. Tour of European Concrete Highways—U.S. TECH* (Washington, D.C., 1992); U.S. Federal Highway Administration, *FHWA Contract Administration Techniques for Quality Enhancement Study Tour (CATQEST)* (Washington, D.C., June 1994).
[14] Stephen C. Lockwood, *The Changing State DOT* (Washington, D.C.: American Association of State Highway and Transportation Officials, 1998).

with considerable community involvement in project design. According to one participant from the design side, neighborhoods were involved to an extraordinary degree, down to the selection of the color of noise barrier materials and the precise location of shrubs and trees in the landscaping.

Of course, such involvement can get out of hand and increase the agency's cost of constructing a facility compared to the old "design and bulldoze" approach. But adherence to that approach in the construction of the urban Interstates led to a backlash and the imposition of community controls because agencies were building projects that were not acceptable to local communities. To some extent, the transportation planning community is still feeling the impacts of that backlash.

The key to effective public participation is to solicit broad enough participation from the community to dilute those with the most extreme views, but not so broad as to bog the process down with too many different actors. Extremists will participate whether or not they are invited. The law provides ample opportunity for them to do so. But the interests of the broader public can easily be lost if only those with extreme views participate. If a project is worthwhile, its benefits are likely to be felt throughout a community. It is important to have representation not only from those who feel the negative impacts, but also from those who enjoy the positive benefits.

Typical tools for outreach and public participation include:

Newsletters: Agencies often develop a mailing list and distribute newsletters to keep interested parties informed about project developments.

Advisory committees: Agencies often provide for one or more committees drawn from the community to advise and act as a sounding board on project design and implementation. The nature and extent of the role of such committees will vary depending on the project. In controversial projects, it may be useful to support an active committee that can serve as a sounding board for project designers and construction managers. The committee might meet regularly (monthly or quarterly) to advise on details of design and construction, such as how best to address citizen complaints about construction noise, traffic, etc. In some cases more than one committee might be useful, organized on the basis of subject area (e.g., landscape, traffic management) or project segment.

Media relations: To the extent a project has a significant impact on a community, the community's news media are likely to cover it regularly or episodically. It stands to reason that friendly, honest, and open access to reporters can help win balanced coverage. On the other hand, if a reporter decides to go for a Pulitzer by attacking a project, it can exact a high cost. In such cases, strong community support, vested in solid advisory committees, can provide some relief.[15]

---

[15] Hagler Bailly Services, Inc. and Morpace International, Inc., *Guidance for Communicating the Economic Impacts of Transportation Investments*, National Cooperative Highway Research Program, Report 436 (Washington, D.C.: National Research Council, Transportation Research Board, 1999).

Monitoring

A final element of flexible planning involves monitoring. Under this somewhat mundane label falls a broad range of critical activities. The purpose of monitoring is to examine how well the system is performing and what threats and opportunities lie on the horizon. To some extent monitoring is similar to strategic planning in that it includes environmental scanning and the development of performance criteria. But strategic planning focuses on the development of goals and objectives to achieve those goals. Monitoring under flexible planning recognizes that the transportation system serves a rich variety of goals: delivering children to day care, delivering the injured and ill by ambulance to emergency rooms, carrying people to and from work, carrying couples to scenic overlooks on starlit evenings. While society might rank one of those purposes as more important than the others, to the travelers involved, each trip is important, and one trip may interfere with the speed or ease of the other.

Performance Indicators

The development of performance criteria is a considerable challenge. The range of performance criteria is large (see Table 9-1). Clearly, many of the performance criteria derive directly from the data categories in Table 9-2. The distinction between descriptive and performance measures is that while descriptive measures are fact based, performance measures include some judgment about whether descriptive measures are adequate. Of course, there is a subjective aspect to the descriptive data as well, since the selection of what to measure and what to make public involves judgments about what users and the public will find useful. But the performance function seeks to go beyond data to assess how well or how poorly the system is functioning.

Interest in performance measurement and performance-based management has risen significantly in recent years. At the federal level, the Government Performance and Results Act of 1993 (GPRA) requires all federal agencies to establish performance measures and goals. Performance measures are one dimension of a broad-based global interest in administrative reform. In the U.S. these efforts were evident in the Clinton-era National Performance Review and Reinventing Government initiatives, as well as in GPRA itself. Public sector reforms of a similar character have recently appeared in Korea, the United Kingdom, New Zealand, Canada, France, Brazil, Australia, and Sweden. A common theme of these reforms is the use of the market as a model for political and administrative relationships, relying on the theories of public choice, principal-agent, and transaction cost economics.[16]

---

[16] "Government Performance and Results Act of 1993," P.L. 103-62, *Stat.* 107 (1993): 285; Linda Kaboolian, "The New Public Management: Challenging the Boundaries of the Management Vs. Administration Debate," *Public Administration Review* 58, no. 3 (May/June 1998): 189-90; Jack H. Nagel, guest editor, "The New Public Management in New Zealand and Beyond," special issue, *Journal of Policy Analysis and Management* 16, no. 3 (Summer 1997); Organization for Economic Cooperation and Development, *Governance in Transition: Public Management Reforms in OECD Countries* (Paris, 1995); Organization for Economic Cooperation and Development, *Public Management Developments: Update 1994* (Paris, 1995).

Table 9-2
Performance Criteria (exemplary)

Traffic counts
Speed
Safety
Energy consumption
Equity
Congestion
Reliability
Continuity
Robustness
Adaptability
Financial viability

One of the central themes of the administrative reform movement is attention to results and performance. For cultural and historic reasons, government agencies have traditionally given great weight to procedural integrity, often at the expense of results and performance. Procurement procedures emphasize fairness and objectivity in making awards, often frustrating the desire of those charged with delivering services for speedy, technically sound implementation. Budget and accounting procedures use "fund accounting," which, unlike the cash and accrual accounting used in the private sector, emphasizes accountability for ensuring that funds are spent at the time and for the purposes for which they were budgeted, but give little attention to results or performance.[17]

Some information about performance can be gleaned simply from comparing descriptive data across jurisdictions. Comparability would require uniformity of definitions and measurement techniques. Such "benchmarking" activities have begun to receive attention. Hartgen began an annual ranking of state departments of transportation in 1992, although some departments have objected to his methodology. More generally, agencies concerned about being compared to other agencies might not wish to comply with any data standards.[18]

Other performance data do not emerge from simple comparison, such as the value produced from the operation of a facility. This difficulty is a direct consequence of the enormous range of goals served by the transportation system. Moreover, the same level of activity might well yield different value depending on the quality of information about, for example, traffic condi-

---

[17] Steven Kelman, *Procurement and Public Management* (Washington, D.C.: AEI Press, 1990); Steven Kelman, "White House-Initiated Management Reform: Implementing Federal Procurement Reform," in *The Managerial Presidency*, ed. James P. Pfiffner, 2d ed. (College Station, Tex.: Texas A&M University Press, 1999), 239-64; Andrew C. Lerner, "Major Components of the Infrastructure Asset-Management Process," last revised, March 6, 1998; presented at the AASHTO/FHWA Workshop on Asset Management, October 1-2, 1997, Rensselaer Polytechnic Institute.

[18] David T. Hartgen and Nicholas J. Lindeman, *The ISTEA Legacy: Comparative Performance of State Highway Systems: 1984-1997* (Charlotte: University of North Carolina at Charlotte, 1999); John Semmens et al., "Improved Method for Measuring Highway Agency Performance," Paper No. 02-2341, in *Innovations in Delivery of Transportation Services*, chair Roy Sparrow, Transportation Research Board 81st Annual Meeting, January 13-17, 2002 Washington, D.C.

tions. With good information, some travelers may choose to reschedule or cancel trips while others might, on the basis of the same information, go ahead and take a trip. Better information yields better decisions, which in turn yield greater value. Observed fluctuations in total miles of travel or total hours of travel might be a poor proxy for total value created by the transportation system. Traffic that increases in the face of better information might actually be producing disproportionately more economic value.[19]

Performance analyses of safety might examine the economic and social impact of fatalities, injuries, and property damage. They would raise questions such as how society should value a life, whether age should be a factor, and whether policies that disproportionately affect children have more or less priority than those that affect the elderly.

Environment performance indicators related to human health, for example, might include not only emission rates, but also exposure and hazard. That is, emissions would describe the mass of emissions (e.g., tons of CO/day), exposure would measure the extent to which people were exposed to the toxin, and hazard would measure the extent to which people's health was adversely affected by exposure to the toxin. Initial progress in the area of full cost pricing has begun to make inroads for these measures. One limitation of the full-cost-pricing literature, however, is that it focuses on costs alone, without consideration of benefits.

Equity measures would examine the level of system accessibility and use by various components of the population. It would break out on a number of dimensions: race and ethnicity, income, location, gender, and age. Descriptive measures would describe access and use by these various subgroups. More evaluative measures could help determine whether current levels of access were sufficient, and other models could examine how system changes might affect such measures. System reliability can be measured through some sort of variance of travel time. Highly reliable systems will have low variances, and less reliable systems will have high variances.

While the heightened interest in performance measurement is encouraging, its full implementation often requires a major cultural shift in an organization or sector, and such shifts can meet with considerable resistance. Important as performance measurement is, the intelligence and decision support elements can sometimes be implemented with less of a cultural shift and with considerable promise of improvement. Thus, intelligence and decision support are flexible planning elements that may be implemented sooner rather than later, whereas full-blown performance-based management may require take considerably longer to put in place.

Monitoring Flexibility

While a set of performance indicators can provide ongoing information about how well a system is performing, some aspects of performance do not lend themselves to continuous meas-

---

[19] D. Brand, "Applying Benefit/Cost Analysis to Identify and Measure the Benefits of Intelligent Transportation Systems," *Transportation Research Record: Journal of the Transportation Research Board* 1651 (1998).

urement. In some cases, developing a specific measurement for a performance dimension is difficult or expensive. In other cases, assessing a particular dimension of performance requires considerable research design. This is especially true of measures of flexibility.

Three of the performance dimensions identified in Table 9-1 relate specifically to flexibility: continuity, robustness, and adaptability. The continuity of a service refers to whether a service once provided will continue to be provided, and whether a facility once provided will continue to be maintained and operated. Notwithstanding the desirability of flexibility, those faced with an investment whose value depends on the continued availability of a service or facility, such as someone whose plant depends on the availability of access to a rail link or Interstate highway, may wish to be confident of continued service provision, or even seek a government guarantee to that effect.

The robustness of performance refers to a system's vulnerability to failures. How vulnerable is a network to the failure of one or more links? How much redundancy is built into the system to allow for continued operation with the failure of particular links or components? The range of failures to be considered might be weighted by a crude assessment of their probability and impact so that more probable or more serious failures would be assessed more frequently, whereas more remote or less damaging failures would be given less attention.

The adaptability of a system refers to its ability to continue to perform under changing "external" circumstances. In some respects, this is the complement to robustness. Whereas robustness assesses performance in the face of failure of a component or subcomponent, adaptability focuses on the ability to perform under changed or unanticipated circumstances.

The word *external* deserves some elaboration. Because there is no firewall between supply and demand, demand is not external to the system per se. But while demand and supply may be simultaneous, demand for transportation facilities and services has not traditionally been under the direct control of facility suppliers and service providers. Demand for transport is derived from demand for other goods and services. Changes in the demand for other goods and services can have important implications for demand for transportation facilities and services. By the same token, changes in transportation facilities and services can modify the demand for other goods and services, as the owners with land adjacent to railway stations and freeway interchanges have long known.

An important element of adaptability is identifying and maintaining options. It involves asking questions about how to maintain system performance in the face of uncertain conditions. In a sense, this is the heart of flexible planning, which seeks to embrace agility over control and stability. It is at the same time the most creative and the most difficult aspect of flexible planning. Whereas traditional planning has sought to identify social goals and use planning to achieve those goals, flexible planning recognizes that social goals are diverse, inconsistent (we want it all and we want it now), and subject to learning.

So rather than imposing control on a chaotic domain in order to achieve agreed upon goals, flexible planning seeks to provide a playing field where households and firms can sort out and resolve the intricate tradeoffs between competing and conflicting goals. This is at once both a market-based and political process. But it is important to recognize that it fastidiously avoids the use of technical solutions to paper over what are at bottom conflicts and rivalries over values. Conflicts over values must be resolved through market and political processes. They cannot be resolved on the grounds of technical criteria. Flexible planning is thus a distinct departure from the Progressive tradition, which seeks the use objective science to find a single best way to solve a problem.

Flexible planning must focus on keeping a range of options open. It should not select which option should carry the day, but allow political and market processes to "fight it out." In this sense, flexible planning is much like the logistics function of the military. It is not clear which weapons or tactics will be needed in a battle situation. The role of logistics is to provide a range of weapons and tactics to commanders.[20]

But flexible planning is more difficult, because there is no single commander or command structure. Rather, it must occur in an arena of conflicting and competing demands. Moreover, the world is full of opportunists who seek advantage through the influence of public authority, for example, to locate a facility near or away from a particular parcel.

The true nature of the "plan" emerges only in retrospect. Shocking as it is to contemplate, we do not know where we are going, we cannot see the future. As the poet said, "The obscurest epoch is to-day."[21] Moreover, society is better off not knowing. If it were known which of today's powerful interests would be eclipsed or disadvantaged by actions being contemplated today, those interests would mobilize to maintain their current power and influence.

James Bryant Quinn, in a classic article, articulates a key process of planning as "consciously preparing to move opportunistically." "Organization and fiscal resources must be built up in advance to exploit [opportunities] as they randomly appear."[22] This expresses exactly the notion of agility. Quinn's analysis derived from the study of successful companies that survived major shifts in products or lines of business. In each case, he found that the ultimate direction of change was not known at the outset.

> [T]he concepts and products that once dominated the company's culture may decline in importance or even disappear. Acknowledging these ultimate consequences to the organization at the beginning ... would clearly be impolitic, even if the manager both desired and could predict the probable new ethos. These factors must be handled adaptively, as opportunities present themselves and as individual leaders and power centers develop....

---

[20] Henry E. Eccles, *Military Concepts and Philosophy* (New Brunswick, N.J.: Rutgers University Press, 1965), 262.
[21] Robert Louis Stevenson, "The Day after Tomorrow," reprinted in *Complete Works* (1924).
[22] James Brian Quinn, "Strategic Change: 'Logical Incrementalism'," reprint, with comment, *Sloan Management Review* Summer 1989: 45-60.

> Strategy deals with the unknowable, not the uncertain. It involves forces of such great number, strength, and combinatory powers that one cannot predict events in a probabilistic sense. Hence logic dictates that one proceed flexibly and experimentally from broad concepts toward specific commitments, making the latter concrete as late as possible in order to narrow the bands of uncertainty and to benefit from the best available information.[23]

Ironically, flexible planning is somewhat antithetical to a long-term view. It rejects the notion that the future is knowable and predictable and that the correct stance is seeking to identify long-term master plans and acting to implement them.

Rather, planning must consciously prepare the transportation system and the society that uses it to move opportunistically, to take advantage of opportunities from technological improvements, scientific discoveries, and other social learning about how best to create value. Planning must build up capabilities and resources in advance in order to exploit opportunities as they randomly appear.

A focus on master planning and the long term is ultimately highly conservative, for it reinforces the status quo. If there is to be a struggle over the selection of a master plan, that struggle will most likely be won by those with the greatest weight and moment today. Directions antithetical to their interests will be avoided, and the power of the state—to compel the payment of taxes, to direct public spending, to condemn land—will be mobilized to protect their interests.

The absence of a purely technically objective basis for master planning undermines the legitimacy of the process and heightens the likelihood that it can become a ritual manipulated by proponents of the status quo—what political scientists call "elites." As a tool of such elites, transportation master planning as practiced today betrays the Progressive ideal that is its foundation.

This betrayal is neither a condemnation of the "road gang" nor of transportation "reformers." Both elites manipulate the process to advance their own agendas.

How then to identify options and select which ones to keep open and which to close off? This is the most difficult question. Highways are the dominant transportation mode, and there is little prospect for closing them down or reducing their role dramatically. They are largely self-financed. The notion of options in highways relates to reserving rights of way for future additions of capacity or filling in the interstices of a network to allow greater interconnectivity or point-to-point route redundancy.

Transit's problems are much different. In most markets, it is a subordinate mode with declining market share. There is little prospect for significantly increasing that share, despite the hopes of many reformers. The notion of options here relates to identifying mechanisms for keeping tran

---

[23] Quinn, "Strategic Change: 'Logical Incrementalism'," 48, 55.

sit available as an option for those that cannot or prefer not to use the highway system. Another option is to consider large-scale deregulation and restructuring to engage the forces of competition to improve service delivery and expand transit markets.[24] There is also value, albeit difficult to quantify, in the preservation of transit capacity as a backup in case of catastrophic failure in the highway system, such as a critical shortage of petroleum or the failure of critical links, as occurred in San Francisco and Los Angeles after their recent earthquakes.

Pedestrian and bicycle capacity are also alternatives that deserve to be maintained and in some cases expanded. There is also value in examining options outside the traditional modal boundaries, looking for hybrids such as paratransit, demand responsive transit, and personalized rapid transit, as well as focusing on the discovery of new modes.

Assessing Financial and Economic Viability

A final aspect of monitoring is assessing financial and economic viability. Providing transportation facilities and services requires funds. The health and well-being of the array of funding sources are critical. While a critical area of planning, the financial side of the issues is also well established already. Most existing operating agencies keep a close eye on their funding so it is not necessary to discuss it in detail here.

The broader economic consequences of transportation are also important. One area that has begun to receive well-deserved increased attention is the assessment of economic impacts above the operating agency level. While agencies have a powerful stake in monitoring their own fiscal and financial condition, the broader economic impacts of transportation facilities and services on economic prosperity, quality of life, and productivity often receive less attention.

Recent initiatives have begun to correct this neglect. Considerable effort has gone into the identification and estimation of total social costs of transportation, including a broad array of "nontraditional" cost dimensions such as environmental impact broadly construed.[25] Also of interest have been efforts to consider the lifecycle cost of facilities, in order to counter the tendency to minimize initial cost at the expense of higher long-term costs.

Consideration of the benefits of transportation facilities and services to the economy is equally important, although it has received somewhat less attention than costs. Traditionally, benefits have been captured largely in terms of reduced travel times. But other benefits are often difficult to quantify and value. Such benefits as reduced total logistics costs, for example, filter through the entire economy but are difficult to capture in a traditional cost–benefit frame-

---

[24] Klein, Moore, and Reja, *Curb Rights: A Foundation for Free Enterprise in Urban Transit*.
[25] Mark A. Delucchi, *The Annualized Social Cost of Motor-Vehicle Use in the United States, 1990-1991: Summary of Theory, Methods, Data and Results*, Report UCD-ITS-RR-96-3 (1) (Davis: University of California, Davis, Institute of Transportation Studies, June 1997).

work.[26] Excessive attention to costs could bias analyses in favor of least-cost investments, even when greater net benefits might materialize with higher costs.

Taken together, these steps can provide a foundation for a reinvention of transportation planning for the twenty-first century. While transportation planning is full of honest, well-meaning people, the process has gotten out of hand. The proposed approach reflects what is beginning to emerge as good practice in some locations and includes suggestions for radical changes. The practicality of implementing such an approach is not self-evident, and that is the question addressed in the next chapter.

***

This chapter has presented a proposal for a new approach to transportation planning in the U.S. This approach seeks to emphasize flexibility and adaptability, through four major steps: intelligence gathering, decision support, design and implementation, and monitoring. Like any proposed new approach, however, its implementation faces a number of potential barriers, and those form the subject of the next chapter.

---

[26] T. R. Lakshmanan and William P. Anderson, "Transportation Infrastructure, Freight Services Sector and Economic Growth," paper presented at the "Importance of Freight Transportation to the Nation's Economy" (Washington, D.C.: U.S. Federal Highway Administration, February 5, 2002); ICF Consulting and HLB Decision-Economics, *Economic Effects of Transportation: The Freight Story*, Final Report, Submitted to the Federal Highway Administration, Under Contract to AECON (January 2002).

10

REALITY CHECK: INSTITUTIONALIZING FLEXIBLE TRANSPORTATION PLANNING

Previous chapters have set forth a proposal for reinventing transportation planning by instituting a flexible approach that is more responsive to changing social and economic considerations. Institutionalizing a flexible planning regime is not going to be easy, however, or without skeptics. In a decision-making environment accustomed to the false certainty of traditional forecasts of traffic, cost, and benefit, a planning approach that emphasizes what is not known is likely to be seen as falling short. Indeed, one federal official literally laughed in the author's face at the suggestion that he tell the congressional patrons of the highway program that the benefits accruing from future highway spending were uncertain. To the federal or state budget official who asks how much to spend on highways or transit, a hard number is essential. Secretaries of transportation and public works directors need a target number with a reasonable rationale if they are to sit down at the table with other agencies and divide up the budget pot.

Moreover, deferring spending on maintenance or new construction has consequences that do not always show up well in a budgetary process dominated by single-year budget cycles and "out years" of four to five years into the future. Turning off spending on maintenance and construction for a year or two or three does have consequences, of course, but they are much less immediate and gut wrenching than turning off spending on, say, school lunches or medical care.

How can flexible planning come into its own in such an environment? If transportation planners have been misguided for fifty years about the nature of decision making, it is not necessarily easy to tell those who pass out the budget that the scales have fallen from your eyes and that you have seen the light.

Going Cold Turkey

One way forward might be to "go cold turkey"—that is, to adopt, painful as it might be, a policy of brutal honesty about what is known and not known about the quality and reliability of the analyses that have traditionally guided decision making, and about the poor quality of the underlying data.

While theoretically appealing, going cold turkey is likely to be unappealing for a variety of reasons. First, it would be a shock to the decision-making system, especially at the federal level. Telling congressmen and senators and Office of Management and Budget officials that the analyses they have been receiving for years have been incomplete and misleading is likely to expose many individuals, both career transportation officials and political appointees, to career-threatening attacks. For that reason alone, this approach is unlikely to fly.

Second, going cold turkey would expose the transportation programs to attack from program opponents and to plundering from other budget areas. While this proposal for flexible planning seeks radical changes in the decision-making process, it does not seek to destroy these programs. So going cold turkey is not an acceptable approach, even if it is intellectually honest.

The remaining alternative is to reform the existing decision-making process. This is much messier, but ultimately is likely to have much more impact

In thinking about how to institutionalize flexible planning, it is useful to keep in mind its four functions: intelligence, decision support, design and implementation, and monitoring. Of these four, intelligence is probably the least threatening to existing institutions. It is not entirely unthreatening, of course, since the power of some institutions resides in access to or control of information. But it is a place to start.

This chapter begins with a brief review of institutional forms that may be available for flexible planning. It then turns to what approaches would be appropriate for instituting flexible planning at the local, regional, state and national levels.

The National Campaign

One approach to mobilizing action is a national campaign in the style of the Interstate highway program of the 1950s and 1960s and the Apollo program of the 1960s. There are two problems with national campaigns, however. First, precipitating such an effort requires a crisis or a national consensus about what needs to be done. Should a crisis occur, it would be wise to be prepared to respond opportunistically. But a wait-for-crisis strategy is not very appealing because it allows current problems to persist indefinitely. Developing a broad-based national consensus in favor of flexible transportation planning, however, would be a difficult sell. Many strong, entrenched constituencies support the current approach, and widespread fundamental change would probably be very difficult to mobilize.

A second difficulty with a national campaign is the risk of what one military strategist has called the "logistic snowball." A logistic snowball is exemplified by the crash program, which can sap other programs of their resources, thereby increasing costs, reducing effectiveness, and leading to requirements for still more resources.[1] Many of the excesses of the early Interstate program are attributable to its status as a crash program intended to complete construction of the whole system in only twelve years. It sapped resources from non-Interstate routes and programs, and its aggressive implementation led to a backlash that remains a powerful influence on highway and transportation policy.

CONSORTIA AND INFORMAL VOLUNTARY ORGANIZATIONS

Since Pressman and Wildavsky's seminal 1973 book *Implementation*, students of public policy have recognized that mandates written in Washington often run aground on the shoals of state and local implementing agencies.[2] "Lower level" organizations on the receiving end of such top-down mandates are adept at circumventing or deflecting them, sometimes by refusing to carry them out, or more often through half-hearted implementation. A federal mandate for a reinvented reformed transportation planning process might founder in the same way.

An alternative to the top down approach of national campaigns is the informal voluntary organization, or consortium, an institutional form that has been emerging in a number of domains, including intelligent transportation systems (ITS). The key feature of such an organization is that it is voluntary and that its formal authority is limited to that which is agreed to by its members. As a result, organizations that join such a consortium give up no authority or autonomy involuntarily. Such organizations have become important mechanisms for coordinating transportation activities in a number of areas throughout the United States.[3]

---

[1] Henry E. Eccles, *Military Concepts and Philosophy* (New Brunswick, N.J.: Rutgers University Press, 1965), chap. 7.

[2] Jeffrey L. Pressman and Aaron Wildavsky, *Implementation: How Great Expectations in Washington Are Dashed in Oakland; or, Why It's Amazing That Federal Programs Work at All, This Being a Saga of the Economic Development Administration as Told by Two Sympathetic Observers Who Seek to Build Morals on a Foundation of Ruined Hopes* (Berkeley: University of California Press, 1973).

[3] This discussion draws on Jonathan L. Gifford and Odd J. Stalebrink, "Remaking Transportation Organizations for the 21st Century: Learning Organizations and the Value of Consortia," *Transportation Research, Part A: Policy & Practice* 36, no. 7 (2002): 645-57. A recent series of reports by Briggs and Jasper provide general guidelines and six case studies of such organization in transportation operations and management. Valerie Briggs and Keith Jasper, *Organizing for Regional Transportation Operations: An Executive Guide*, tech. rept. no. FHWA-OP-01-137 (Washington, D.C.: U.S. Federal Highway Administration, August 2001); idem, *Organizing for Regional Transportation Operations: Arizona AZTech*, tech. rept. no. FHWA-OP-01-140 (Washington, D.C.: U.S. Federal Highway Administration, August 2001); idem, *Organizing for Regional Transportation Operations: San Francisco Bay Area*, tech. rept. no. FHWA-OP-01-142 (Washington, D.C.: U.S. Federal Highway Administration, August 2001); idem, *Organizing for Regional Transportation Operations: New York/New Jersey/Connecticut TRANSCOM*, tech. rept. no. FHWA-OP-01-138 (Washington, D.C.: U.S. Federal Highway Administration, August 2001); idem, *Organizing for Regional Transportation Operations: Southern California ITS Priority Corridor*, tech. rept. no. FHWA-OP-01-141 (Washington, D.C.: U.S. Federal Highway Administration, August 2001); idem, *Organizing for Regional Transportation Operations: Vancouver TransLink*, tech. rept. no. FHWA-OP-01-143 (Washington, D.C.: U.S. Federal Highway Administration, August 2001); idem, *Organizing for Regional Transportation Operations: Houston TranStar*, tech. rept. no. FHWA-OP-01-139 (Washington, D.C.: U.S. Federal Highway Administration, August 2001). For a discussion of similar organizations' roles in responding to concerns about global climate change, see Francis X. Neumann, Jr., "Organizational Responses to Complex and

In the private sector, consortia have gained considerable attention over the past decade as a means for gaining competitive advantage and keeping up with rapid industry change. For example, in the computer and telecommunication industries Hewlett-Packard participates in several hundred standards consortia worldwide.[4]

A consortium can be defined as an agreement, combination, or group of organizations formed to undertake an enterprise beyond the resources of any one member. The rationale behind such arrangements can be explained in terms of transaction cost economics: if the value of two firms together is greater than their value separately, it may be useful to form a partnership.[5]

A partnership can take many forms, but at its core it involves some kind of bonding that holds the agreement together. This bonding can take at least the following three forms or a combination of them:

1. a contractual agreement directing each partner's responsibilities and rights (included here are licensing agreements, joint ventures, etc.);
2. a self-enforcing agreement in which the cooperation is so natural that no contract is needed; and
3. a hostage arrangement in which each partner invests a valuable asset that has to be forfeited if it leaves the partnership.

For the implementation of flexible transportation planning, the *voluntary* versions of consortia are particularly interesting. Using the above partnership typology a voluntary consortium could then, at the extreme, be described as a self-enforcing joint venture with no hostages. However, in reality such extreme versions of consortia are rare. This strong definition emphasizes that voluntary consortia are characterized by having very low entry and exit barriers, because no assets are held hostage to the agreement.

Consortia are not panaceas, and at least three considerations may limit their effectiveness. First, consortia may operate more effectively in environments with a high degree of trust (high-trust societies). This notion has been thoroughly discussed by Francis Fukuyama in his book *Trust*. He defines trust as "the expectation that arises within a community of regular, honest, and cooperative behavior, based on commonly shared norms, on the part of other members of that community." Hence, people in a high-trust society expect other people to fulfill their obligations. In a low-trust society, the risk of being exposed to free riders and contract breakers is assumed to be higher. Lack of trust must therefore be replaced by formal contracts, which increase transaction costs.[6] Therefore, a society with a high degree of trust allows its participants

---

Changing Environments in the Great Lakes Basin: The Adaptability of Michigan State Agencies to Conditions of Global Warming" (Ph.D. diss., Department of Public & International Affairs, George Mason University, 1995).
[4] Brian D. Unter, "The Importance of Standards to Hewlett-Packard's Competitive Business Strategy," *ASTM Standardization News*, December 1996, 13-17.
[5] Paul H. Rubin, *Managing Business Transactions: Controlling the Cost of Coordinating, Communicating, and Decision Making* (New York: Free Press, 1990).
[6] Francis Fukuyama, *Trust: Social Virtues and the Creation of Prosperity* (New York: Free Press, 1995).

to interact with less friction. This suggests that voluntary consortia would be less effective in low-trust societies.

Second, the desirable organization of an industry involves a tradeoff between flexibility and efficiency. Organizations that are efficient in a static environment may be inefficient in a dynamic environment, and organizational structures that are inefficient in a static environment may be efficient in a dynamic environment.[7] Finally, during periods of rapid structural change in an industry, coalitions, consortia, and alliances may be more common. They are more easily dissolved than the are new internal divisions developed or mergers consummated, their sunk costs are lower, their commitments are less irreversible, and their inertia lower.[8]

Two thumbnail case studies from the domain of intelligent transportation systems illustrate the consortium concept. Both have been successful, although in different ways. The first focuses on information sharing and facility management. The second focuses on developing technical standards.

TRANSCOM

TRANSCOM (Transportation Operations Coordinating Committee) is a coalition of seventeen transportation and public safety agencies in the greater New York City metropolitan area, which encompasses parts of New Jersey and Connecticut and has a population of 19 million.[9] Member agencies include state highway departments and law enforcement and emergency response agencies. TRANSCOM, established in 1986, grew out of an effort among agencies responsible for operating the highways in the region to share information about major incidents whose effects spilled over to other jurisdictions, such as major construction projects, accidents, and facility closures. It initially employed low-tech equipment: telephones, pagers, and faxes. It has now evolved into a 24-hour 7-day-a-week communications center with a staff of twenty-nine that coordinates incident response and facility operations and management in the region. In 1998 it incorporated as a nonprofit corporation, TRANSCOM, Inc.[10]

E-ZPass

Another example of an informal voluntary organization is the Interagency Group (IAG), which developed the regionwide electronic toll tag called "E-ZPass" for the greater New York region. To avoid the proliferation of incompatible tags, eight toll agencies in the greater New York re-

---

[7] B. H. Klein, *Dynamic Economics* (Cambridge: Harvard University Press, 1987).
[8] C. U. Ciborra, "Innovation, Networks and Organizational Learning," in *The Economics of Information Networks*, ed. C. Antonelli (Elsevier Science, 1992), 93.
[9] U.S. Bureau of the Census, *State and Metropolitan Area Data Book, 1997–98*, table B-1.
[10] Briggs and Jasper, *Organizing for Regional Transportation Operations: New York/New Jersey/Connecticut TRANSCOM*; Louis G. Neudorff and Tom Batz, "A Regional ITS Architecture for the New York Metropolitan Area," In *Intelligent Transportation: Realizing the Future*, Third Annual World Congress on Intelligent Transport Systems, Orlando, Florida (Washington, DC: ITS America, 1996), 191-92; Frank J. Wilson, "The TRANSCOM Coalition: Multi-Jurisdictional Issues in ITS," *ITS Quarterly* 4, no. 2 (Spring 1996): 83-96. See also TRANSCOM, www.xcm.org.

gion began to meet in late 1989 to discuss development of an interoperable electronic tag. In 1990, they organized themselves as the IAG. Of particular concern was the risk that tags from different agencies might destructively interact with each other. Another concern was that adoption of one technology by a large agency might preempt the options of the smaller agencies.[11]

The IAG issued a request for proposals in January 1992. After a two-stage procurement they awarded a contract for Mark IV tags in March 1994. The pass was placed into operation in October 1995. As of early 2002, it had expanded to sixteen operating agencies in seven states, with approximately 6.8 million tags in use and 3.9 million separate accounts (many households and businesses have a single account for multiple tags). Ancillary use of the tag for roadside purchases has also begun to occur. A McDonald's in New Jersey has begun accepting E-ZPass as a form of payment, and some airport parking concessions accept it.[12]

Of considerable importance in the development of E-ZPass was the need to have eight independent agencies, each with an independent procurement process, jointly procure a tag-reader system that would be technically interoperable, so that one tag would work at all facilities in the system. The technical tradeoffs were significant. Closed toll systems (i.e., where tolls are collected when exiting the system based on the point of entry) required read–write technology. However, open toll systems (i.e., with tolls collected at plazas at strategic points on the main lines) could have used read-only technology, which at that time was considerably cheaper than read–write. Yet through the actions of the IAG the agencies were able to decide on a read–write tag that would satisfy all of the users' requirements, even though it imposed higher costs on agencies with less demanding requirements.

Interoperability of user accounts posed considerable challenges. Ideally, a user would maintain a single account for one or more tags and use that account to pay tolls at any participating facility. The eight participating agencies viewed user account interoperability as a greater challenge and potentially a source of considerable delay. Its implementation was therefore delayed until later.

The New York jurisdictions later entered into a contract with Lockheed Martin IMS for a clearinghouse, and the New Jersey and Delaware jurisdictions contracted with a consortium of Chase Manhattan Bank and MFS Network Technologies. The Port Authority of New York and New Jersey initially signed on to the Lockheed contract with the understanding that it would shift to the New Jersey contract once that was in place. As of early 2002, such account interoperability is currently in place across all participating jurisdictions.

---

[11] This section draws from J. L. Gifford, L. Yermack, and C. Owens, "The Development of the E-ZPass Specification in New York, New Jersey and Pennsylvania: A Case Study of Institutional and Organizational Issues," Second World Congress on Intelligent Transport Systems (Yokohama, Japan, November 9-11, 1995), 1420-26; idem, "E-ZPass: A Case Study of Institutional and Organizational Issues in Technology Standards Development," *Transportation Research Record* 1537 (November 1996): 10-14.

[12] New Jersey Turnpike Authority, personal communication (February 28, 2002); "New E-ZPass Uses to Include Airport Parking," *New York Times*, December 3, 2000; Kathleen Carroll, "Put That Burger on My E-ZPass," *New York Times*, May 27, 2001, 3: 2.

According to participants, a key feature of the IAG was its voluntary, informal organization. Because participation was nonmandatory, agencies were able to arrive at consensus on a workable solution, which would have been much more difficult to achieve in a nonvoluntary or coercive situation. If federal officials had attempted to impose the read–write requirement on these agencies, for example, the outcome could well have been delay, resistance, and political brinkmanship. With the IAG, however, member agencies were willing to sacrifice their own interests in the interest of greater benefits for the region at large.

Finally, of importance for the initial development of the IAG was the fact that the leaders of the member organizations knew each other from IBTTA (International Bridge, Tunnel and Turnpike Association). This personal familiarity allowed them to work together for more than four years from the initial meeting of the consortium until the agencies started signing a formal memorandum of understanding (MOU).

These thumbnail case studies suggest that consortia, or informal voluntary organizations, may be a useful tool for implementing flexible transportation planning. Consortia provide a forum for identifying areas of possible consensus, tradeoffs, and respective interests. While not panaceas, these organizations may be serve as a useful tool for implementing change.

THE DATA SHARING MODEL

One of the central elements of the proposed flexible planning approach is the sharing of information across institutional boundaries. It plays a key part in all three phases: intelligence, decision support, design and implementation, and monitoring. The impetus to share data arises from at least two sources. First is the cost of collecting data using traditional survey methodologies. Second, information sharing hopefully leads to better decisions.

Since the 1920s, state highway departments have cooperated in the collection of data on highway conditions for needs studies. Municipalities and metropolitan planning organizations have cooperated extensively in the collection of travel surveys and inventories needed for transportation planning. Beginning in the 1960s, the Federal Highway Administration (FHWA) in cooperation with the states, devised a continuous highway reporting system. Since 1978, FHWA has sponsored the Highway Performance Monitoring System (HPMS), which contains data on a statistical sample of highway sections in each state, as well as some so-called areawide statistics at the metropolitan and state level.[13]

An increasingly powerful impetus for data sharing is the growing quantity and quality of real-time data resulting from improvements in information, sensor, and communication technologies. "[T]he sharing of information and the attendant decentralization of decision-making in

---

[13] Norman C. Mueller, "Fifteen Years of HPMS Partnership. Accomplishments and Future Directions," <www.tfhrd.gov/pubrds/summer95/p95su10.htm>, *Public Roads* 59, no. 1 (Summer 1995); Kevin Heanue, *Data Sharing and Data Partnerships for Highways*, National Cooperative Highway Research Program, Synthesis of Highway Practice, no. 288 (Washington, D.C.: National Academy Press, 2000), 4.

system operations have given rise to a host of fascinating and hitherto unexplored theoretical and practical problems. These problems center on approaches for optimizing large-scale distributed systems in which participants act as semi-autonomous agents who share some goals (such as taking maximum advantage of available system capacity and resources) while, at the same time, also are in competition with one another."[14]

The problems and opportunities arising out of data sharing are several. First, the effectiveness and efficiency of data sharing is subject to the cost and accuracy of any necessary translation of data from one party to another. If parties are exchanging data using standardized definitions, these transaction costs can be fairly low. If data elements are not standardized, their exchange can be costly and time consuming, and data integrity may be sacrificed if errors occur.

Additionally, standardizing data elements can be an extremely time-consuming process involving the coordination of dozens or hundreds of separate jurisdictions, many with different information requirements. Federal grant agencies like FHWA and the Federal Transportation Administration (FTA) have a strong interest in having uniform data from grant recipients because this facilitates "rolling up" data to the regional and national level. States and localities, however, often have unique local conditions, legacy systems, and institutional arrangements that may or may not complement federal requirements. This is nowhere more evident than in the area of geographic information systems (GIS).[15]

One effective approach to data sharing has been developed with great success in the area of environmental regulation. The Environmental Protection Division of the Georgia Department of Natural Resources for years had wrestled with the U.S. Environmental Protection Agency (EPA) over the quality and timeliness of its mandated reports on the state's environmental matters. EPA claimed the state withheld required data deliberately, or at best neglected its reporting requirements. The state held that EPA rarely used the data the state submitted, and when it did often failed to incorporate updates and changes from the state, so that the state and federal agencies often had sharply different data from essentially the same data source.

After years of tension, the state agency became a pioneering model for a radically different kind of cooperative data sharing agreement with its federal counterpart. Under the new arrangement, the federal government provided computing capacity to host the state's operating databases. Federal EPA pledged to observe the state's ownership and stewardship of the data, and never to change data in the state's files. The state, for its part, opened its files to EPA so that federal EPA could directly query state data for purposes of analysis and reporting. Later, the state also made its data available to regulated businesses and the public.

The state–federal data sharing agreement became a model that was implemented nationally. Moreover, at the national level, EPA launched its own initiatives to make data available to in-

---

[14] Cynthia Barnhart et al., "Workshop on Planning, Design, Management and Control of Transportation Systems," A Report to the National Science Foundation, NSF award #9729252, June 15, 1998, 3.
[15] Jack Faucett Associates, COMSIS Corporation, and Mid-Ohio Regional Planning Commission, *Multimodal Transportation Planning Data*, Final Report, Project NCHRP 8-32(5), Prepared for National Highway Cooperative Research Program (Transportation Research Board, National Research Council, March 1997).

terested parties. For example, the Toxics Release Inventory is a public database of company filings of their toxic chemical releases, available through the National Library of Medicine.[16]

Such a state–federal data sharing approach deserves serious consideration in transportation, at least for many data requirements. It preserves the autonomy and accountability of the states and substate jurisdictions, while meeting data and reporting requirements at the national level. Data sharing and data partnerships for the collection of HPMS data exist in a few states (California, Michigan, and Pennsylvania). In these cases, a metropolitan planning organization (MPO) or local government participates in the collection of HPMS data and, in some cases, is reimbursed for its costs through the Federal-Aid Highway Program.[17]

A LOCAL, STATE, AND REGIONAL IMPLEMENTATION STRATEGY

How would one go about establishing and institutionalizing a flexible planning regime at the local and regional level? This would need to vary from city to city, depending on local institutions and history. Several important questions arise. What is the scale of the planning institution? Should it correspond to the jurisdiction of existing localities, or should it be regional?

Consider the functions of the flexible planning regime: intelligence, decision support, design and implementation, and monitoring. From the standpoint of consumers of intelligence, it would be helpful to have some uniformity of definition and presentation in data that would allow cross-jurisdictional comparisons, rolling up data to the regional level, and so forth. This suggests that a regional institution might be useful. The same could be said of decision support, design and implementation, and monitoring. Yet the search for regional organizations of appropriate scope and authority in transportation and in other fields will be a very long one.[18]

Metropolitan Planning Organizations

One option is to vest responsibility for flexible planning in the MPOs that operate in more than 200 of the largest U.S. metropolitan areas. Such an approach might make sense where MPOs are well developed and have become institutionalized. Many MPOs, however, do not meet this test. The development of regional planning in the U.S. has faced a number of barriers, chief among which have been efforts among local jurisdictions to preserve their autonomy. While MPOs have been established by federal fiat since the 1960s as a condition for obtaining federal highway funds, many have been kept weak by local and state institutions that have not wished to cede authority to them.

---

[16] These initiatives are documented in four case studies, Malcolm K. Sparrow, "The State/EPA Data Management Program," parts A–D (Cambridge, Mass.: Kennedy School of Government, Harvard University, 1990), 2-3.
[17] Heanue, *Data Sharing and Data Partnerships for Highways*.
[18] Jonathan L. Gifford and Danilo Pelletiere, "'New' Regional Transportation Organizations: Old Problem, New Wrinkle?" CD ROM paper no. 02-4009, Transportation Research Board, 81st Annual Meeting (Washington, D.C., 2002).

Federal policy, especially the Intermodal Surface Transportation Efficiency Act of 1991 (ISTEA), has sought to strengthen MPOs by giving them veto power over capital spending programs. But they have also saddled them with the responsibility for developing and enforcing plans to attain conformity with air quality standards, which is politically very challenging in many areas. MPOs are thus being forced by federal policy to promulgate plans that are arbitrary and bear little relation to any objective assessment of future prospects. With that as their responsibility, their credibility as a source of objective, factual intelligence and decision support is problematic.

Moreover, in many ways the model of the MPO as a decision-making framework is exactly what flexible planning seeks to displace. Metropolitan planning organizations are artifacts of a top-down, nationally uniform decision-making process. They are the stewards of the modeling and decision-making process that is fundamentally flawed. Whether one should reform an MPO or try instead to displace it will depend greatly on the particularities of the MPO in question.

Informal Voluntary Organizations

An alternative to relying on existing MPOs to institutionalize flexible planning is the creation of an informal voluntary organization. As discussed earlier, these organizations have emerged as mechanisms for coordinating transportation activities.

Vesting authority for flexible planning in such an organization might begin with a metropolitan conference, convened by a university or other neutral party or leading organization in the region, that invites from all the regions major decision makers with transportation briefs. The purpose of the conference would be to explain flexible transportation planning and explore its appropriateness for a particular region.[19] Depending on the outcome of such a conference, the organization could start small, simply developing mechanisms for sharing data among themselves and with the public at about the level and scope of existing operations. This could be as simple as a web site. That is, the effort would start with the development of an intelligence function. Later efforts would concentrate on decision support and the other functions of flexible planning.

The central feature of such an approach is *not* that it would provide a forum for resolving conflicts over values. That is a problem for political and market institutions. What such a consortium would do, however, is provide a forum for identifying threats and opportunities, areas of consensus over how to resolve them and for acting on them quickly once they emerge. In areas where there is considerable disagreement, conflict will not magically be resolved through such a proc-

---

[19] This idea was initially developed by students in a course at George Mason University, who developed it as a means of establishing a regional consortium of transportation operating agencies in the Washington metropolitan region. Jonathan L. Gifford, Matthew Hardy, and C. Owens, "Voluntary Organizations in the Deployment of Intelligent Transportation Systems: Case Studies and Applications to the Washington, DC Metropolitan Region," ms. (1996).

ess, but that is as it should be, since our system of government is organized to move slowly in the absence of consensus.[20]

There is a certain irony in vesting flexible, non-goal-oriented planning in a voluntary institution with no formal authority. And there is a risk that the consortium could turn out to be impotent, incapable of doing anything, like many councils and commissions.

Yet such an approach would seem to be the only hope in many areas. Authority to initiate and execute projects is nailed down tightly by existing agencies, as are authority and rights to veto or oppose such projects. There is little prospect of simply steamrolling such opposition with regional government. While a consortium might eventually evolve into a regional authority of some sort, it is unlikely to start out that way. Existing authorities and their constituents would forcefully oppose such a move.

The consortium approach, while imperfect and uncertain, provides a path for development of a level of regional cooperation that can work in the context of a region's existing institutions and history. What works in the Washington, D.C., metropolitan area may well not work in Seattle or San Francisco. But the notion of a uniform regional planning and decision-making framework has been tried for the last thirty-five years in the form of MPOs. And the evidence is not at all compelling that it has led to improved decision making except in a few cities.

Given the limited success of vesting authority in MPOs, it makes sense to try something new. Where it works it will provide a foundation for decision making. Where it fails—and it will not work everywhere—it may be no worse than continued pursuit of the status quo, with its halting, litigious rituals.

How might such a system work in the realm of the intelligence function of flexible transportation planning? Many agencies collect a broad range of data as part of their routine operations. Some of this data, such as bus maintenance records, is fairly arcane and of little public interest. Other data, such as traffic counts, traffic accident and fatality incidence, bus schedules and on-time performance, air quality readings, and parking violation citation locations, holds broader interest. With the advent of the World Wide Web and the Internet, the hardware cost of making such data available has fallen to less than $5,000, well within the equipment budgets of most agencies.

The Role of the States

States can play a critical role in the adoption of flexible planning. As the owners and operators of state highway systems, they have an enormous stake in system performance. They can adopt the principles of flexible planning for their state highway and transportation systems, and they can facilitate their adoption by substate units through grant incentives, enabling legislation, or direct subsidies for activities to initiate flexible planning.

---

[20] I am indebted to my colleague Brien Benson for articulating this point.

Because states are major suppliers of highways and transportation facilities in many metropolitan areas, they are active participants in the local transportation scene. Indeed, state departments of transportation were the principal agents for the construction and operation of the Interstate system in all major American cities. As major participants, active endorsement from states could significantly further flexible planning. States might, for example, serve as administrative hosts for intelligence gathering and information dissemination. Or they might work through state universities to support such activities. Because states are often also interested parties in many transportation policy disputes, care must be taken to avoid real and apparent conflicts of interest.

Another important role for the states would be facilitating cooperation across state boundaries in cases of multi-state metropolitan areas, of which there are many in the U.S. They can establish multi-state entities such as the Port Authority of New York and New Jersey. Where states are unwilling or unable to be actively supportive, a policy of neutrality towards the adoption of flexible planning by substate units would be helpful.

FLEXIBLE PLANNING AT THE FEDERAL LEVEL

The introduction of flexible planning at the federal level is in many ways more complex than its introduction at the state and local level. Much of that complexity derives from the complex web of programs and interests that operate at the federal level, both within the transportation sector and across the full range of government functions. It would be naïve to expect a wholesale restructuring of national transportation programs to reflect new and untried principles of flexible planning, however logically sound. What steps at the federal level would be appropriate for beginning to test and refine these principles?

Identify and Address Uncertainty in Program Plans and Analyses

All government agencies face significant uncertainties that affect their ability to manage and implement their programs. These range from uncertainty about which party will be in power in the executive branch and Congress to uncertainty about future economic conditions, climate, technology, and public opinion.

Many program areas incorporate uncertainty directly into their planning and programs. The recognition of uncertainty is particularly evident in budgeting and economics. The Treasury Department must recognize and manage uncertainty about interest rates, currency exchange rates, economic growth, and inflation. The Office of Management and Budget and the Congressional Budget Office recognize that the budgetary impacts of a program or legislative change are uncertain, and re-estimate budgets routinely to take advantage of better information.

Unlike many federal programs, highways have traditionally relied heavily on twenty-year forecasts. As discussed extensively in this book, the objective foundation for much of that twenty-year analysis is highly questionable. While this is relatively well accepted in the technical community, policy makers have sought to broaden the scope of the highway-oriented approach to include transit and other modes of transportation. Thus, beginning in 1993, the U.S. Department of Transportation's (DOT) biennial report on conditions and performance (C&P) was expanded to include transit.[21]

Extending this planning regime to transit and other transportation modes moves in the wrong direction, however, away from a more defensible posture and towards reliance on unsupportable long-term forecasts of intrinsically unpredictable conditions. It does provide consistent information across transportation modes, and therefore greater intermodal comparability. But extending the false characterization of certainty from highways to other modes is anything but encouraging.

The U.S. DOT did introduce an improved methodology to the C&P reports beginning in 1997. While the agency continues to rely on problematic traffic forecasts for the estimation of user benefits, it does focus on the economic contribution of surface transportation improvements to the economy.

The difficulty lies in responsibly presenting uncertainty to decision makers. Members of Congress and their staffs want a defensible number. The question they pose to program administrators is "What level of spending is needed for the next budget period?" They want a technically accurate number, but they do not want to be numbed by the details and assumptions that go into that getting to that number.

Nonetheless, incorporating uncertainty into the economic analysis is essential. As time horizons for estimating the benefits of projects expand, so does uncertainty about the economic value of those benefits. Estimated timesavings in the fifteenth or twentieth year of a project are highly speculative, and that uncertainty should be incorporated into the budgeting process and systematically disclosed.

A further consideration is the intense pressure at the federal level to reduce expenditures and remain within spending caps. While budget surpluses, when they occur, relax this pressure somewhat, an economic reversal or downturn can revive budget deficits. Moreover, the funding requirements of politically untouchable programs like Social Security and Medicare are making significant and growing claims on budgets.

Budgets for capital improvements and maintenance can suffer in such circumstances because the costs of deferred maintenance and capital improvements only become evident over the course of many years. The Highway Trust Fund has been a bulwark against competing budgetary priorities in years past. But in recent decades the trust fund has come under increasing pressure, both for expenditures for nonhighway purposes, first transit and more recently "enhance-

---

[21] U.S. Federal Highway Administration and Federal Transit Administration, *Condition and Performance: 199 Status of the Nation's Surface Transportation System*, Report to Congress (Washington, D.C., 1997), xxi.

ment" improvements to landscape, historic buildings, parks and recreational areas, and other categories that do not provide direct transportation services.

These concerns have given rise to a range of suggested policy remedies. Some have suggested a capital budget at the federal level. Others have sought to move the Highway Trust Fund off budget.[22]

In this environment of intense budgetary competition, for program advocates, open acknowledgement of uncertainty would be unthinkably poor politics. Program critics or those competing for budget resources would welcome any admission of uncertainty about program needs.

Acknowledgment of uncertainty need not play into program reductions, however. An alternative way to tell the story is this:

1. Highway maintenance and capital expansion have powerful impacts on the cost of goods and services and on economic competitiveness.
2. The U.S. is blessed with an efficient, low cost transportation system.
3. An important contributor to low costs and high efficiency has been the nation's highway program, in which the federal government plays a central role.
4. Every billion dollars of federal spending on highways translates into x billion dollars of tangible lower costs for goods and services for American families and businesses, or y dollars for every American household.
5. Moreover, highway improvements offer considerable other benefits, such as improving safety and allowing families to have access to more choices in where they shop, where they worship, and what schools they attend. Businesses use an efficient transportation to improve productivity, cut costs, and compete more effectively in the global marketplace.
6. Highway spending is subject to diminishing returns. Spending the entire budget on it would not be sound. But at current and even significantly increased program levels, reductions in tangible costs outweigh program costs by a factor of z.

The economy is too complex, too dynamic to know the exact level of spending that will maximize social welfare. Moreover, it is subject to uncertainties about the effectiveness of planning and administration, contingent actions by other parties (like business co-investment), and chance events like oil price increases and earthquakes. Moreover, money spent on unused or underused roads will always hurt the economy.

But intelligent spending on highways contributes to economic productivity and competitiveness. The work of Nadiri, Aschauer, McGuire, Munnell, and others, though not yielding precise estimates of economic payoffs, has suggested that the "external" benefits of infrastructure

---

[22] Moving the trust fund off-budget was one of the most contentious issues in the 1998 reauthorization.

are measurable at the national level, even if they are difficult to identify at the level of individual companies and households.[23]

Because such an argument is intellectually honest, it can stand up to challenges to its assumptions, and its assumptions can be tested and evaluated. The key difficulty of such an argument, of course, is getting reasonable estimates for x, the cost reductions per billion dollars of federal program spending, as a function of program level. Such data are not available today, but the Federal Highway Administration (FHWA) of the U.S. DOT could certainly begin to develop them, using a combination of internally available data and contracting out a set of forecasts an economic consulting firm. Indeed, FHWA has begun to sponsor efforts to estimate how the transportation system affects the performance of the freight sector.[24]

Shorten the Time Horizon for Project Analysis

As noted earlier, federal procedures rely heavily on twenty-year forecasts for highway traffic as a basic metric for program benefits,. The federal planning requirements should move to focus more on the projects that pay off in a five- to ten-year horizon. This would serve to concentrate investment on expenditures with a nearer term payoff.

The incorporation of economic investment rules beginning in the 1997 conditions and performance report goes a long way towards reducing the importance of highly uncertain long-term benefits, simply because at most discount rates, benefits beyond ten to fifteen years are discounted substantially from their real dollar value. At a discount rate of 10 percent, benefits are reduced to 39 percent after ten years and to 15 percent after twenty years. At a 5 percent discount rate, the corresponding reductions are 61 percent and 38 percent, respectively.

Economic discounting alone may still give considerable weight to highly uncertain long-term benefits, however. Present values may derive a considerable fraction of their total net benefits between the tenth and twentieth year. For a level benefit stream over twenty years, 27 percent of the present value accrues from the tenth to twentieth years at a 10 percent discount rate. At a 5 percent discount rate, the corresponding figure is 38 percent. If benefits are assumed to rise continuously, an even greater fraction of present value would be attributable to the "out" years.[25]

---

[23] For recent literature surveys, see Marlon G. Boarnet, "Highways and Economic Productivity: Interpreting Recent Evidence," *Journal of Planning Literature* 11, no. 4 (May 1997): 476-86; and T. R. Lakshmanan and William P. Anderson, "Transportation Infrastructure, Freight Services Sector and Economic Growth," paper presented at "The Importance of Freight Transportation to the Nation's Economy" (Washington, D.C.: U.S. Federal Highway Administration, February 5, 2002).

[24] Reports from FHWA's freight benefit–cost analysis study are available at <http://www.ops.fhwa.dot.gov/freight>.

[25] These estimates use continuous compounding, so that $DPV = \int_{T_1}^{T_2} Be^{-it} = B\left(\frac{e^{-iT_1} - e^{-iT_2}}{i}\right)$, where DPV is discounted present value, i is discount rate, and T_1 and T_2 are the beginning and ending time periods. See F. M. Scherer, *The (Julia?) Child's Guide to e*, case no. N16-92-1127.0 (Cambridge, Mass.: Harvard University, Kennedy School of Government, 1992).

The federal government could begin to move towards flexible planning by adopting a shorter term for project analysis. Alternatively, it could require a "risk premium" on discount rates for projects so as to reduce further the contribution of long-term benefits, which are subject to considerably greater uncertainty.

Adopt Incentives Based on Measurable Outcomes

Perhaps the most important reform necessary at the federal level is to move away from the mandated plans designed to achieve environmental targets defined in Washington. These are reminiscent of the worst of the Soviet and Chinese central planning charades, where Moscow and Beijing set the national production targets and all "subordinate" units doctored their production figures to show compliance. It is the antithesis of rational, objective decision making.

The result of such a policy is that policy makers at all levels are constrained from giving their best and most objective assessments of what is achievable and what policy objectives are reasonable. U.S. DOT and FHWA officials repeatedly state that they cannot generate reports or present analyses that show that localities will not comply with the requirements of the Clean Air Act and ISTEA, even though many of them will privately say that their best estimates indicate that may well be the case.

This is not to say that the federal government should abjure governance of air quality or other policy initiatives. Far better than forcing localities to produce unachievable forecasts would be using federal incentives based on demonstrated, verifiable achievements. Such an approach in air quality policy, for example, would withhold or grant highway funds on the basis of observed emissions, air quality, or compliance with the National Ambient Air Quality Standards (NAAQS), rather than on the basis of plans or promises for achieving them.

Such an approach could have the added benefit of making the definitions of conformity with the NAAQS more useful than the definitions employed today. Today's determinations of conformity tend to be highly sensitive to variations in local weather conditions. A particularly warm summer or series of summers can throw a metropolitan area out of conformity, and a string of cooler summers can take them into conformity, with no variations in levels of emissions.

The question of how to specify incentives so that they are effective is complicated, in air quality governance and elsewhere. In the case of air quality, should the emphasis be on the policy initiatives that state and local governments can actually adopt, such as inspection and maintenance programs? Or should incentives be based on plans to adopt policy initiatives, which may induce some action on the part of stakeholders but which may ultimately never be implemented?

California's electric vehicle requirements are a case in point. In 1990, California mandated that 2 percent of vehicle sales in 1998 had to be zero emission vehicles (ZEVs). The minimum percentage rose to 5 percent in 2001, and to 10 percent in 2003. ZEVs were expected at that time to be powered primarily by electric batteries. Other states found California's requirements appealing and adopted them. The Clean Air Act Amendments of 1990 permitted states to adopt the California standard as one option for attaining conformity, which several northeastern states, including New York and Massachusetts, chose.

In February 1996, California relaxed its ZEV requirement by entering into a memorandum of agreement (MOA) with auto producers that retained the 10percent requirement for 2003, but eliminated numerical targets until 2003. In November 1998, California relaxed its ZEV requirement for 2003 to 4 percent from 10 percent, provided auto manufacturers sell enough of a new category of "super ultra low emission vehicles" (SULFVs). In response to California's relaxation in 1996, vehicle manufacturers in New York and Massachusetts successfully sued those states, arguing that the Clean Air Amendments of 1990 did not allow standards more stringent than California's.[26]

Part of what motivated this policy brinkmanship was the need to have a plan in place to meet the NAAQS. California was able to buy time in its quest for conformity, but it did so by imposing considerable regulatory uncertainty on auto manufacturers about what targets would be in place for which years.

When federal incentives are based on promised action and anticipated instead of actual results, this tends to lead to the manipulation of promises that may ultimately be unattainable, but that impose considerable inconvenience on business, industry, and households. There may be a place for technology-forcing regulation, but it brings with it many problems.

Another approach to measuring effectiveness is focusing on the level of emissions measured in the local area, over which governments have less direct control but which is closer to the policy objective of cleaner air. Still another approach is to focus on the measured level of toxics in the air, such as carbon monoxide. Measured levels of toxics depend not only on emissions but also on ambient climate conditions, which are beyond the control of local and state governments but are closer to the objective of protecting human health. Or the focus could be on human health effects of air pollution, which is one of the major policy objectives but which is difficult to control from a policy perspective and difficult to measure precisely over time.

The point is not to resolve this problem here, since selecting the appropriate policy measure for air quality incentives is complex and requires considerably more study and analysis. It is to illustrate that incentives based on promises or adopted plans should generally be avoided because they tend to invite gamesmanship among those governed by the incentives and because they tend to complicate and obfuscate policy analysis and assessment at the federal level.

---

[26] California Air Resources Board, "ZEV Fact Sheet: Memoranda of Agreement" (September 8, 1998) <www.arb.ca.gov/msprog/zevprog/moa.htm>; "Realistic Goals: A Revised Electric-Car Mandate," *San Diego Union–Tribune* (July 9, 1998); PR Newswire, "Market-Based Approach to Zero Emission Vehicle Program Working for Californians" (July 30, 1998); PR Newswire, "Air Board Continues California's World Leadership in Auto Emission Standards" (November 6, 1998); Dow Jones News Service, "Fedl Appeals Ct Throws Out N.Y. State Electric Cars Rules" (August 11, 1998); and Ellen Perlman, "Breakdown in the ZEV Lane," *Governing* (November 1998), 66.

Facilitate Sharing of Information

In addition to adopting program incentives based on measurable outcomes, the federal government could support flexible planning by facilitating the sharing of information among affected communities and stakeholders. To some extent, the Bureau of Transportation Statistics (BTS) has already initiated this process. BTS was established in 1991 under the authority of ISTEA and reauthorized in the Transportation Equity Act for the 21st Century (TEA-21).

U.S. DOT has begun several initiatives along these lines. In December 1998, FHWA completed its reassessment of the Highway Performance Monitoring System (HPMS), one of its basic data systems.[27] The National Cooperative Highway Research Program (NCHRP) sponsored a project on multimodal transportation planning data aimed at assessing and improving data resources for planning.[28] In addition, NCHRP has a number of ongoing projects under its research problem area 8-32. It is also preparing a synthesis of highway practice on data sharing and data partnerships for highways (synthesis topic 30-07).

These efforts at exploring and fostering data integration, cooperation, and sharing can provide a valuable foundation for states, localities, and other operating agencies that seek to enhance the quality of their data.

In conclusion, implementing a flexible planning approach in the *realpolitik* of contemporary transportation decision making is a formidable challenge. Going "cold turkey," while intellectually honest, is unlikely to have much impact. Mobilizing a national campaign for reform is a worthy initiative, but also a formidable effort. Yet there are practical, intellectually honest steps that localities, states, and the federal government can take that will begin to break the hold of traditional methods that are widely recognized to be inadequate. Taken together, initiatives at the local, state, regional, and federal level to develop voluntary consortia and shared data can help explore the utility and practicality of adopting flexible urban transportation planning. Consortia and voluntary organizations can provide valuable opportunities for organizations to explore new working relationships without compromising their autonomy or their constituents' interests.

While the challenge of a flexible approach to planning is considerable, the need for it is urgent. The traditional approach has deteriorated into an empty ritual that is itself very costly and does little to conserve precious resources—natural, social, and economic—and support a competitive society that provides opportunities to all for achieving a better life.

***

This chapter has provided a "reality check" on the proposed flexible approach to transportation planning. In particular, it has examined how evolution to the proposed approach could occur in

---

[27] U.S. Department of Transportation. Federal Highway Administration, *Highway Performance Monitoring System Reassessment*, final report (Washington, D.C., December 1998).

[28] Jack Faucett Associates, COMSIS Corporation and Mid-Ohio Regional Planning Commission, *Multimodal Transportation Planning Data*, final report, prepared for National Highway Cooperative Research Program (Transportation Research Board, National Research Council, March 1997).

light of current institutions, interests, resources, and capabilities. In particular, it has examined how consortia and voluntary organizations could provide a mechanism for implementing the proposed flexible approach. The next chapter provides a step-by-step agenda for adopting a flexible approach to transportation planning for cities, counties, states and the U.S. government.

11

AN AGENDA FOR ACTION

What actions are necessary to begin to institute change in the current system? This chapter presents an action agenda for those wishing to foster flexible transportation planning. Surface transportation problems and opportunities always have a local impact, and sometimes a regional or larger impact. So this agenda includes a strong local element. Any agency or community can tailor the principles of flexible planning to its own activities. State highway departments, transit operators, metropolitan planning organizations, and city and county transportation agencies can each institute an intelligence function that makes their data available to interested parties via the Internet or other media.

The agenda laid out below is a menu of actions that companies, chambers of commerce, interest groups, agencies, and other interested parties can initiate to move beyond the expensive, frustrating stalemate that characterizes decision making in many areas today. Of course, Portland's solutions are unlikely to work for Detroit, and each region will need to tailor the action agenda to its own local conditions, history, institutional setting, and economic standing.

LOCAL, STATE, AND REGIONAL ACTIONS

An effort to reform transportation decision making can start at the local, regional, or state levels. Institutions at these levels play key roles in decisions about the location, design, and timing of transportation facility development. They also have access to the local knowledge and conditions that are central to striking the balance between community and environmental stewardship, and transportation facility development. This section outlines steps to initiate such an effort.

Establishment of an Intelligence Function

A key first step towards implementing a more flexible approach to urban transportation planning is establishing a factual basis for discussion—provided by an intelligence function. The appropriate mechanisms for doing so will depend on local conditions. The following are options for getting the ball rolling.

Convene a regional conference

The key to initiating change is convening the interested parties. This is a considerable task, especially in light of the highly conflicted nature of transportation decision making in most U.S. cities. Moreover, some of the important stakeholders might see any change from the status quo as a threat, either to themselves or to their employing agency.

The identity of the conference convener is therefore important. The convener would preferably be a neutral party. A university or research institute might host the conference. Where a neutral party is unavailable, two or more organizations representing a range of perspectives might act as co-hosts.

For that reason, it would be important for the conference to focus on the exchange of information as a first step. While this agenda may seem to be innocuous, many agencies may be reluctant to go even that far. Agencies that have been pilloried or sued might be unwilling to release their internal data because of the potential for inciting controversy. Transportation is a hot-button issue in many metropolitan areas. An emphasis on data relating to present and historical conditions, however, may be sufficient to enlist participation, at least by many agencies.

On the other side, interest groups that have sued agencies to stop or modify projects may be loath to sit down at the same table, especially if lawsuits have been or continue to be acrimonious. Again, a focus on strictly verifiable data should facilitate cooperation.

Those interested in facilitating change may wish to move slowly at first, convening a small working group initially and moving later to a regional conference. However, the initial group should be balanced and wary of appearing biased towards or against a single set of solutions.

Inventory stakeholders

It may be useful to inventory transportation stakeholders in the region. Recognize that some stakeholders may opt out, at least at the beginning. Remain open to their joining later.

Identify basic descriptive indicators

Identify a set of indicators that can characterize the dimensions of the local transportation system of interest to current and future stakeholders. These indicators should include many of those mentioned in Chapter 8. They should adequately describe all modes, including, to the extent possible, bicycle and pedestrian.

Capture and make available data about the region

Inventory data stores available in the region, and make them available, perhaps only in raw form initially. The inventory could include historic reports and surveys. It could be combined with a collection in a public or university library that could house hard copies of historic reports that cannot easily be made available electronically.

Based on the descriptive indicators identified above, canvass the regional transportation operating and planning agencies to identify data sources. Of particular use would be data gathered as a byproduct of operating activities, which would reduce the costs of collecting data.

While a set of compatible data and data definitions would be useful at some point, its development requires considerable effort, including significant changes to legacy systems and procedures. As a first step, simply make data available along with their definitions. This would provide stakeholders in the community with the opportunity to learn and use what is there. Later, a set of regionwide compatible data definitions for a limited number of priority measures could be developed, and agencies could be coaxed to modify their systems to produce data accordingly. The effort required to achieve such compatibility is considerable and should not be underestimated. Requiring understaffed agencies to modify their processes and procedures will usually not foster participation and cooperation.

Develop a Decision Support Function

While some communities will elect to develop an intelligence function and no more, others may wish to go further and develop a decision support function. Any decision support function is likely to require some ongoing commitment of human resources for system development, maintenance, and technical support.

Develop summaries and extrapolations of raw data

The simplest decision support tools would summarize and extrapolate current and historical data collected as part of the intelligence function. (Indeed, the boundary between the two steps is not crisp.) Such summaries can be useful for demonstrating a community's current and historical conditions. Short- to medium-term extrapolations (three to five years) can help a community understand the near-term implications of the continuation of current trends.

Inventory current models and data

Communities might also want to use the extensive models and data that comprise the federally prescribed urban transportation planning procedures. For medium to large metropolitan areas, the designated metropolitan planning organization (MPO) would be a good place to start. These agencies are required to develop and maintain regional plans as a condition for receiving federal transportation funds. They often have an infrastructure of data and models that may be useful for decision support.

Use of federally sanctioned tools and institutions imposes a number of limitations, however. The models typically used for federally prescribed transportation planning procedures are complex and require considerable expertise to operate and understand. Moreover, many MPOs are short of staff and may not be able to support a grassroots enterprise, which they might also see as a threat. Some MPOs are highly political and sensitive to anything that could cast them or any of their constituent jurisdictions in an unflattering light. Under the right circumstances, however, MPOs could be extremely useful participants.

Identify a host or hosts for the decision support facility

A neutral or trusted host for the decision support function and its associated tools and data could be a local planning agency, a university, or a newly created (or existing) consortium. Each community will determine what is feasible and appropriate for itself, depending on the types of problems it is experiencing and the available resources. The decision support facility would, at a minimum, host the data sets and tools, presumably in a web-accessible format. Additional functions could include technical support, meeting space, dispute resolution, and group facilitation, if resources are available.

Investigate the feasibility of GIS

Geographic information systems (GIS) are powerful tools for organizing and managing spatially oriented data such as transportation data. They have also become increasingly user-friendly. Their implementation at a regional level, however, poses some significant hurdles because they require a high level of data integration. That is, to be imported into GIS, data sets must use a consistent set of geographical references; thus much of the raw byproduct data that are readily available may not be easily imported into the GIS.

Design and Implementation

One of the most important parts of the transportation planning and development process, of course, is the design and construction of facilities. Indeed, in the "here and now" of a particular facility in a specific place in a specific community, advocates and stakeholders from different perspectives can actually sit down and strike some compromises about how to resolve safety, mobility, community preservation, and environmental impact in an anticipated project.

The prime movers on projects are implementing agencies: departments of public works and transportation at the state and local level. An independent flexible planning organization can be made available as a resource to work with implementing agencies to involve stakeholders in planning and decision making. Indeed, such agencies typically seek the participation of community groups as part of their community outreach and public participation activities. A flexible planning organization can serve as a neutral forum to mediate conflicts. Many agencies will welcome the participation of a broad-based group as opposed to the more common single-issue neighborhood or environmental group.

Conduct Monitoring

Monitoring the performance of the urban transportation system and providing that information to the public is an important function of flexible planning. It can foster accountability and focus the attention of the public and their public officials on problems and opportunities.

Develop performance indicators

Above and beyond the neutral descriptive data described above are performance indicators. The flexible planning organization should try to design a consensus set of indicators, and then marshal the data and resources to develop and maintain them. While developing such indicators would be a difficult step beyond providing descriptive statistics, such indicators can function as a powerful tool for holding operating agencies and political bodies accountable. A consensus set of indicators will necessarily be fairly broad. Chapter 9 presented a sample list of indicators.

Conduct flexibility and viability reviews

More ambitious monitoring activities would include review and assessment of the flexibility and financial and economic viability of the current transportation system. Such reviews would be a significant undertaking and would probably not be feasible until the flexible planning organization had been in existence for several years.

NATIONAL ACTIONS

For the last half century, the federal government has played a central role in defining and institutionalizing the current urban transportation planning process. Beginning with the Interstate highway program, the federal highway program sharply increased the coerciveness of its policies. It shifted away from the "associative philosophy" of its early years, which relied on technical leadership and encouragement of good practice and sound procedure. Of course, the federal government could play an important role in reforming that process.

Deregulate the Metropolitan Transportation Planning Process

In the transportation planning process, the federal government is inappropriately placed squarely in the middle of states' and localities' decisions about how to adjust their local transportation systems to shape and accommodate the mobility of households and firms. While there is clearly a national interest in a transportation system that accommodates interstate commerce, public health and safety, and national defense, the federal influence is out of scale to its interests. It relies on a command-and-control approach to regional, top-down metropolitan planning. This approach, required by regulation in every large metropolitan area, has left little room for experimentation and the discovery of alternative techniques and approaches.

Revise Federal Policies to Focus on Measurable Outcomes

Federal policy, particularly air quality policy, relies heavily on promises instead of results. Much of the bureaucratic apparatus for implementing these policies revolves around who participates in the development of such promises (plans) and then trying to force localities to deliver on their promises.

A significant improvement would be to revise such policies to focus on results, not promises. Air quality policy, for example, might allocate some portion of federal highway funds on the basis of actual air quality or improvements in air quality. While this may be seen as a radical suggestion, reorienting policy from promises to results would likely lead to a more rapid improvement in air quality and more creative efforts from states and localities in trying to achieve better results. Moreover, the data documenting results could be made publicly available and accessible to the community.

To get the ball rolling, Congress, the Executive Branch, or private groups could commission studies to ascertain how such a results-oriented policy could be implemented, canvassing worldwide to ascertain if such policies have been implemented elsewhere.

A results-oriented approach might be useful in areas other than air quality policy. Safety funds could be allocated on the basis of safety improvements. Transit funds could be distributed on the basis of ridership.

Streamline the Environmental Approval Process

The current environmental approval process introduces extraordinary delays into project development. These delays severely hamper the ability of the transportation system to respond agilely to changes in social and economic conditions.

Such streamlining need not sacrifice environmental integrity. Indeed, the current process may be unintentionally increasing the scale of projects. Planners routinely expect a new project to require a decade or more from conception to completion, and hence expand the scale of the facilities they propose in anticipation of traffic growth in the interim.

A streamlined process would permit smaller, more frequent improvements to the transportation system. Lessons learned about the impacts of earlier projects could be used to tailor the designs of future projects to maximize usefulness and minimize environmental impact.

The 1998 reauthorization of the surface transportation programs (the Transportation Equity Act for the 21st Century, TEA-21) called for the U.S. Department of Transportation and the Environmental Protection Agency to develop an environmental streamlining proposal, which is, as of this writing, in process.[1] But streamlining of the approval process is an urgent priority.

Shorten the Time for Project Cost–Benefit Analysis

The federal government should shorten the required planning horizon for project analysis from twenty to ten years. In conjunction with the streamlining mentioned above, such shortening would significantly increase economic returns from the transportation program. While facilities generally have a service life beyond ten years, they also become increasingly obsolete as they age. While ten years, like twenty, is an arbitrary horizon, it is preferable for several reasons. First, ten-year forecasts are likely to be much more reliable than twenty. Second, given the pace of technological and structural changes occurring in the economy, from telecommuting to intelligent transportation systems (ITS), the highways we build today are often obsolete before they are completed. *We are building the legacy systems of the future!* It is a favor to future generations to be as parsimonious as possible in constructing facilities, building only what is clearly needed in the near to medium term and leaving open for them options to expand further.

Such a recommendation flies in the face of the recent trend towards life-cycle cost analysis (LCCA), which recommends analysis horizons of up to forty years, and the Federal Highway Administration's policy statement on LCCA, which requires a horizon of at least thirty-five years.[2] The intent of such policies is to counter the tendency towards lowest initial cost as the analysis criterion for maintenance and rehabilitation. Such planning horizons may be appropriate in circumstances where conditions appear to be fairly predictable, such as pavement management. But they generally overlook the option value of waiting to invest, which can be considerable. In cases where conditions are unpredictable, such as many major infrastructure investments, a more incremental approach using shorter planning horizons is more appropriate.

---

[1] U.S. Department of Transportation, *Highway and Transit Environmental Streamlining Progress Summary*, Report to Congress (February 2002).
[2] James Walls III and Michael R. Smith, *Life-Cycle Cost Analysis in Pavement Design: Interim Technical Bulletin*, U.S. Federal Highway Administration, FHWA-SA-98-079 (Washington, D.C., September 1998), xii.

Support a National or Regional Data Interchange Standards Process

The intelligence function of local- and state-level units would be considerably enhanced if data from different groups were compatible. Such compatibility would allow different groups to exchange and compare data with each other. It would also allow the aggregation of data across groups, which would be helpful for state, regional, and national officials. A further benefit of data standards would be the interoperability of software that utilizes such data.

It is essential, however, that any such standards be voluntary. If adherence to such standards became a condition of eligibility for funds from the Federal-Aid Highway Program, for example, most communities would comply. Such a mandate, however, risks sabotaging a cooperative intelligence function. While standards for data might seem uncontroversial, groups and communities that value attributes of the transportation system differently could easily disagree about measurement. Whether and how to measure bicycle and pedestrian traffic, for example, could be extremely controversial, as could methods for measuring airborne emissions, noise, and other environmental impacts of transportation. While relying on voluntary adoption of standards will be a slow and arduous process, resorting to mandates is likely to prompt controversy and dissension.

One advantage of a voluntary approach is that it is likely to yield a broad range of descriptive measures. Furthermore, as conditions change and give rise to the need for new and different measures, a voluntary process can provide a great deal of flexibility in developing measures. While a mandated standard might appear appealing in the first instance, it is often difficult for a regulatory process to keep abreast of changing conditions. As researchers identify new things that need to be tracked, a voluntary system can respond much more readily than a mandated regulatory process.

Such a standards-setting body might take a number of institutional forms. To some extent, existing standards-setting bodies are engaged in activities that overlap considerably with the compilation of data for transportation intelligence activities. Mapping data, for example, has been extensively treated in various standards-setting bodies for purposes of vehicle navigation systems and GIS. Working through existing standards bodies would take advantage of past investigations and findings. A transportation working group or technical committee could help prioritize areas for standards development. Procedures for doing so are well established.

Alternatively, transportation intelligence functions might be institutionalized in an independent body through umbrella organization such as the National League of Cities, the National Association of Regional Councils, the Association of Metropolitan Planning Organizations, and the National Governors Association. One disadvantage of such an approach, however, is that the parent organization generally represents one constituency and might be seen by other constituencies as not completely objective.

A third alternative is to create a standards consortium to represent the interests of flexible transportation planning organizations in the deliberations of the standards-setting bodies. Since the cost of participating in standards-setting processes (travel, preparation, and meetings) can be considerable, individual localities may not find participation feasible. One source of technical support and expertise could be the Bureau of Transportation Statistics (BTS), which has made considerable strides in making transportation data available since its creation in 1991.

Reorient National Planning around Macroeconomic Analysis

National highway planning is now entrenched in the "conditions and performance" (C&P) paradigm. The methodology used in the reports issued under this paradigm has been improving and now takes into account user costs and the cost–benefit ratios of projects under consideration. These are considerable improvements over the engineering standards (or indicative planning) approach of previous years.

However, the C&P paradigm continues to have some serious weaknesses as a policy and planning tool. It is subject to extraordinary data limitations due to the nature of the Highway Performance Monitoring System (HPMS), especially in forecasts of future traffic, which, in turn, drive much of the economic analysis. Moreover, the core economic model, the Highway Economic Requirements System (HERS), estimates economic returns on fairly simplistic assumptions about how states program highway funds.

The national planning effort would be much more useful if it were reoriented towards results, on the one hand, and macroeconomic modeling predictions, on the other. A results orientation would lead to efforts to estimate *actual* returns from investments in prior periods. The emerging emphasis on transportation asset management, both at the federal and state levels, would dovetail nicely with such a reorientation. So too would the requirement by the Government Accounting Standards Board (GASB) that governments state the value of their infrastructure assets on their balance sheets and recognize depreciation of infrastructure on their income statements.[3] Under this approach, reports to Congress would emphasize the results the program was producing—the rates of return on investment—rather than on relying on rather flimsy assertions from the C&P reports about program levels required to maintain or improve system performance.

In addition to a results-oriented reporting system, it would be useful to develop a macroeconomic forecast of demand for transportation five to ten years into the future, broken out at the regional level. While the predictability of such models is inherently uncertain, they could provide useful input to Congress in determining appropriate program levels.

***

This agenda for action at the local, state, regional, and national level charts a course for moving away from the increasingly stylized and ritualistic transportation planning process that governs most metropolitan regions today. The resources invested in metropolitan transportation are

---

[3] Governmental Accounting Standards Board, *Basic Financial Statements—and Management's Discussion and Analysis—for State and Local Governments*, Statement No. 34, Governmental Accounting Standards Series, no. 171-A (Norwalk, Conn., June 1999).

enormous—hundreds of billions of dollars per year—as are the positive and negative impacts they impose on the economy. But the project cycle has become so protracted that society's ability to refine and improve the transportation system to support increasingly complex and stringent social, economic, and environmental requirements is seriously impaired. The process is badly broken. This agenda for action seeks to chart a way forward that will help society discover the proper balance between infrastructure improvements, changes in lifestyle and behavior, and stewardship of financial and fiscal resources for future generations.

12

CONCLUSION

Urban transportation in the United States is at a critical juncture. In hundreds of metropolitan areas around the country, growth in road traffic has outpaced growth in road capacity. The resulting congestion delays travelers and raises fuel consumption and air and noise pollution. Traffic congestion also increases logistics costs for goods distribution, which leads to higher prices for consumers and misallocation of capital assets in the economy.

One solution to this problem is the adoption of correct pricing strategies, which would lead to economically efficient utilization of available capacity. The benefits of such a strategy appear to be great enough to compensate those made worse off by road user charges, with benefits to spare. While promising, such strategies have not been widely adopted in the U.S. or elsewhere, and they face a number of political and implementation barriers.

Another set of solutions would manage demand, in particular redirect growing demand for road space by private automobiles and light trucks to public transit and shared vehicles. Public transit and shared vehicles have an important but tiny market share in the U.S. Although growing in recent years, this market promises to remain tiny for the foreseeable future.[1]

This book has focused on the urban transportation planning process, the decision process that has given rise to the current dilemma. The author has questioned the basic assumptions and practices of transportation planning as it is practiced in the United States today. He has suggested the adoption of a more flexible approach.

A central pillar of modern transportation planning is what is often thought of as a systems approach. This approach adopts a synoptic view of the transportation system at a metropolitan or regional level. It examines how changes in endogenous and exogenous factors will affect a

---

[1] Anthony Downs, "How Real Are Transit Gains? While Transit Advocates Suggest There's Been a Massive Shift in Travel Behavior, Those Claims Appear to Be Exaggerated," *Governing* (March 2002): 54.

vector of performance variables over twenty years. For practical reasons, it suppresses analysis of localized (within-zone) movements of pedestrians, bicycles, and automobiles.

This perspective fails to capture the highly dynamic micro- and meso-scopic processes of land use and travel characteristic of a vital metropolitan area. The focus on aggregates that can be modeled successfully at the macroscopic level biases decisions about how to adjust the infrastructure to accommodate changes in demand to support medium- to long-distance travel, when the lion's share of travel is short and local. This much is recognized, if not universally accepted.

Current efforts to improve transportation planning emphasize the development of computer models that predict more accurately micro- and meso-scopic processes (i.e., micro- and meso-simulation and agent-based models). These are promising endeavors, but the application of these models has largely been limited to traffic simulation, and the challenges of implementing them to analyze travel behavior—that is, the decisions that individuals make about where, when, and how to travel—remain daunting.

Moreover, micro- and meso-simulation risk missing an important weakness of the current approach: that it seeks to base decisions on long-term predictions of human behavior and capability that are inherently uncertain over twenty years or longer. Refining long-term forecasting and modeling tools is certainly important. But modifying decision processes to recognize inherent uncertainties is of equal or greater importance.

The current transportation planning regime is a product of the Interstate highway program, which dominated U.S. road transportation in the half century after the Great Depression. The Interstate was a political compromise between a system intended to serve local urban travel and interregional travel. Federal highway planners of the late 1930s saw urban travel as the central challenge and opportunity following the rural highway programs of the 1920s and 1930s. Yet President Franklin Roosevelt, Congress, and many states eschewed addressing local traffic problems and favored developing a national system of high-speed superhighways for long-distance travel. The resulting 1956 Interstate highway program was a hybrid that included the national system and its extensions into and around cities, designed for long-distance intercity traffic, with high speed road geometry and grade-separated interchanges. States and localities were to be responsible for highways and transit to carry local trips. The system received dedicated funding through the Highway Trust Fund and was to be completed in a crash program of only twelve years.

Alas, it did not work out that way. Like most crash programs, it sapped the resources of other programs. With priority funding paying 90 percent of the cost, states and localities opted to build urban Interstates over local projects, both transit and highway. Lacking alternatives, local traffic flooded the new Interstates. Highway planners, alarmed at the rapid traffic increases on the new highways, expanded the scale of subsequent projects, sometimes to ten or twelve lanes. The impact on communities increased commensurately, compounded by the cost-saving practice of locating Interstates in parks, along waterfronts, and through low-income neighborhoods. The freeway revolt was born.

Interstate highways are an inefficient way to carry local traffic, however. Boulevards, arterials, and parkways provide designers with considerably more ability to minimize community and environmental impacts and are much more suitable for short trips. Expensive Interstate engineering features—long vertical and horizontal curves and long acceleration and deceleration lanes—increase impacts dramatically but produce little benefit at moderate speeds. What good is a highway designed to carry traffic at 70 mph when drivers are sitting in gridlock trying to go 10 miles to work or shop?

As a result, many cities are stuck with a freeway network that is not well designed for local traffic and an entrenched opposition that sees Interstate-style highways as public enemy number one. The backlash against the urban Interstates has led many to conclude that highways are the problem, that more highways only lead to more travel and congestion, and that new highways are an inappropriate response to today's increasing road congestion. This is a mistake. While the Interstate facilitated an enormous increase in highway mobility during the 1960s and 1970s, its excesses have also largely precluded the addition of much more highway capacity in many U.S. cities since that time.

Cities all over America are struggling to discover how best to address the legacy of the Interstate highway program. The "new urbanists" have it partly right. Out-of-scale highways have had a devastating impact on many American cities, especially those that grew in the pre–World War II era. The freeway revolt of the 1970s rightly brought many projects to a halt. Yet increased congestion and limited mobility will not be solved by doing nothing. And solutions like improved transit and congestion pricing have limited, albeit important, roles to play.

The physical highways are one of the most tangible legacies of the Interstate program. But another legacy of comparable importance lies in the processes and procedures that govern urban transportation planning. The Interstate-era planning requirements locked all major metropolitan areas in America into a uniform, one-size-fits-all transportation planning process, which remains largely in force today.

This book seeks to rethink urban transportation planning by throwing out much of the canon of policy and practice of the last half century and fostering a transportation system development process that is agile and responsive to changes in social values, preferences, knowledge, and technological capabilities. This new process must recognize that decades of economic prosperity is feeding a strong, deeply seated desire by many for the privacy, security, flexibility, and convenience of private vehicles and the lifestyle they afford.

The answer is not to force people into high-rises and buses—although those options should remain available for those who choose them—but to design communities that are both people- *and* car-friendly, and to design highways that are community friendly.

230 *FLEXIBLE URBAN TRANSPORTATION*

But designing community-friendly highways today is an extraordinary challenge. It is especially challenging in built-up areas that have reserved little room for new moderate speed boulevards, parkways, and expressways, and with a public highly skeptical of promises that road and highway expansion will mitigate congestion.

This book argues that solutions will not arise out of grand battles between competing visions of alternative futures of car-free, car-limited, or highway friendly communities. Solutions will arise instead through collective action, one neighborhood and one community at a time. To engage those communities, it is imperative to shift the discussion away from debates about competing long-term visions of the future and focus instead on near- to medium-term solutions to traffic and mobility problems. And the transportation supply community must recognize that polishing twenty-year forecasts misses the point.

For the last three decades, the public has been in a state of denial, favoring transit "for the other guy" because road-based solutions appear to be so unattractive. The anti-highway community has done an impressive job of casting the automobile and the highway as the enemy of a positive, sustainable future. Their supporters have made almost axiomatic the assertion that "you can't build your way out of congestion."[2] And the solutions being advanced by the highway community often do not meet the test of public acceptability. But if more Interstate-style highways are not the answer, what is?

Unfortunately, the right set of solutions is not known or knowable today. Communities are not faced with a well-defined set of mutually exclusive alternative futures from which they can simply choose on the basis of a well-calibrated multi-attribute decision model. Instead, they face a great deal of uncertainty about their current situation, and profound uncertainty about future conditions and capabilities.

The only way to discover such solutions is at a grassroots and community level. A one-size-fits-all federal planning process simply will not work. What is needed is to liberate communities from the federally proscribed planning procedures that have governed transportation planning for the last four decades and allow them to discover new ways to solve their problems.

Such liberation is more difficult than it sounds. Highway opponents have inserted checkpoints and legal hurdles in the traditional planning process. Inadequate as the current process may be from a technical and logical standpoint, it is the status quo. And planning agencies may open themselves to the uncertainty and controversy of legal challenges if they try to deviate from those procedures.

As congestion increases, so will its political salience. The highway community needs to avoid retreating into a position of saying "We told you so" and offering up recycled, already-rejected proposals. The reason that the reform community has been so successful in portraying

---

[2] James A. Dunn, *Driving Forces: The Automobile, Its Enemies, and the Politics of Mobility* (Washington, D.C.: Brookings Institution Press, 1998).

highways as the locus of environmental evil is that the public *is* concerned about the community and environmental impacts of Interstate-style highways. They want other solutions that more carefully balance improved service with environmental and community preservation.

Some reforms are appearing on the horizon. Portland has sought to utilize an urban growth boundary, with mixed results. St. Louis is attempting to develop a planning process that relies more on community inputs to identify and prioritize problems, rather than on measures of engineering efficiency alone. Large-scale regional plans are less common today, and communities are placing greater reliance on corridor-level studies. And context-sensitive highway design has begun to gain ground as a design approach, at least in some sensitive areas.

This book has sought to showcase and generalize these and other developments, and to provide a portfolio of ideas and approaches available to communities to confront their transportation problems and craft creative solutions to them. It is based on five principles. The first is the urgency of being honest about what is known and not known about current and future conditions. Far too much effort in transportation planning is spent debating long-term forecasts that have little if any relevance to current problems. Indeed, that is one reason it is often so difficult to engage any but those who are passionately committed or professionally obligated to participate in the planning process. Honesty is also essential in discussing the realistic prospects options for addressing transportation problems. In most cases, that is going to include the private automobile as a major part of the solution. While bicycling, walking, and transit should play an important role in transportation, the private auto must as well. It is the predominant form of personal transportation in almost every place outside of New York City.

Second, the foundation of any planning effort must be intelligence, that is, data and information about current and historical conditions to inform participants about where they stand today and where trends appear to be leading. This does not require complex analytical and forecasting models, but rather should be based on raw data and neutral trend extrapolation over a five- to seven-year period. Data should be widely available, preferably over the Internet or successor technologies.

Third, flexible transportation planning can provide decision support in the form of more sophisticated extrapolation and simulation tools. Such resources might be housed in a university or other neutral setting, available for interested parties from all points of view. Such tools could assist communities in coming to decisions and making tradeoffs between infrastructure enhancements and expansions, and the stewardship of financial, community and environmental resources.

Fourth, flexible planning should provide a forum for assisting with design and implementation. When consensus on an investment is reached, flexible planning requires broad community participation in the detailed design of facilities and their construction or implementation. Many jurisdictions practice such outreach activities now, and they play an important role in ensuring that facility improvements are acceptable to affected communities.

Finally, a flexible planning program should have a monitoring element that develops and maintains performance indicators, monitors options and opportunities, and assesses financial, economic, and environmental viability. Such an activity might be housed in a university or, more likely, in a public or private regional planning organization.

Flexible planning tries to pick what is best about emerging reformed planning practices and combine it with important insights about flexibility and agility. The transportation profession is building the "legacy systems" of the future. It has a profound responsibility to preserve options and opportunities for future generations. But the profession must also recognize its obligations to the present generation, not only to the well off, but also to the less fortunate, less capable, and less economically endowed.

Current policy is heading towards a train wreck. The notion that public policy can force society to give up mobility and single-family, large-lot living and opt for higher densities and use of transit, bicycles, and walking is fanciful—even punitive. Public preferences for the automobile will almost surely win out in the end. The next decade or two could be spent in a holy war between highways and transit, or it could be spent in building and developing flexible urban transportation planning processes that will yield facilities and communities that more closely reflect the values and preferences of their residents. This is the challenge.

To date, much of the highway community has viewed the opposition as elitist, seeking to force-feed society a set of lifestyle and policy prescriptions that are out of touch with mainstream American preferences. Some of the extreme environmental proposals fit this characterization. But the reason that the highway opposition has gained such influence in the political and policy-making arena is that the public is concerned about the community and environmental impacts of highway construction and development. While few among the general public are willing to get out of their cars and into buses or carpools to act on their concerns, they are concerned enough to support policies and politicians who seek to address environmental and community impacts.

The highway community must wake up to these concerns and embrace environmental integrity and community preservation as legitimate social objectives, not simply as regulatory and procedural hurdles. If the highway community can develop project proposals that address mobility problems *and* enhance the environment and the community, they can mobilize mainstream support for modifying and improving infrastructure so that communities work better, now and in the future. That is the challenge, and the enormous opportunity.

The City of Calgary, Transportation
Phone: (403) 268-1174
Fax: (403) 268-1874
email: donna.pickard@calgary.ca
ISC: Protected

- options mandated - common
  monitor
- maintenance & design & implementation
- public support - strategies & evaluation tools
- improved service in environmental & community recognition
- Honesty - known about current & future conditions, realistic
  movement
- funded & options for existing transp. system problem
- Data - current & historical conditions
- decision support ?
- balance of programs & community needs
- solutions at programs & community level
- tear to short term solution to tough 8 yr file, 10+
- even in projects which project in 5 to 10 yrs.

232 Public Policy is important - are positive

gopark of Ring Rd ; PIC → policy ;

Jivraj, Azim

From: Pickard, Donna
Sent: 2009 July 07 8:56 AM
To: Pickard, Donna; Atkins, Dianne; Bolger, John; Cataford, Anne; Jivraj, Azim; Kok, Ekke; Vanderputten, Ryan; Jordan, Chris; Blaschuk, Chris; Wong, Calvin; Berting, Randi; MacNaughton, Eric
Cc: Mulligan, Don
Subject: RE: COMPULSORY PRE-READ - PLAN IT CALGARY SESSION - JULY 23

A little reminder to keep passing the books along!

Donna Pickard
Executive Assistant
Transportation Planning, #8124
The City of Calgary, Transportation
Phone: (403) 268-1174
Fax: (403) 268-1874
email: donna.pickard@calgary.ca
ISC: Protected

From: Pickard, Donna
Sent: 2009 June 17 8:08 AM
To: Atkins, Dianne; Bolger, John; Cataford, Anne; Jivraj, Azim; Kok, Ekke; Vanderputten, Ryan; Jordan, Chris; Blaschuk, Chris; Wong, Calvin; Berting, Randi; MacNaughton, Eric
Cc: Mulligan, Don
Subject: COMPULSORY PRE-READ - PLAN IT CALGARY SESSION - JULY 23
Importance: High

There are four copies of a book titled "Flexible Urban Transportation" by author Jonathan L. Gifford in circulation to all those attending the "Plan It Calgary Implementation" meeting scheduled for July 23rd, Fort Calgary.

Please ensure that you read Chapters 8 through 12 (Chapters 1 through 7 provide a history overview of transportation planning) and use the information to prepare a one-page 'point form' summary to answer the following:

1. What is your interpretation of the information you have read in Chapters 8 through 12; and
2. What are the implications for Transportation Planning and the implementation of Plan It Calgary going forward?

Please bring your summary to the July 23rd meeting for discussion.

Thanks.
Donna Pickard
Executive Assistant
Transportation Planning, #8124

2009/07/10

10/07/2009

160 RTN forecast 15 years, Interstate model, Values, environmental integrity, Values = environmental integrity, vitality
168 Fig 8-1
176 Focus on near-term problems & opportunities, decisions by household, firm and
177 community definitions
178 Debate about values
180 Truth from previous pl.
181 Congestion
182 Data - intelligence function
184 Co-operative effort w/ local univ.
185 - And Impact rethinking - model Ching
- Design & procurement - prof/cons
187 - Public participation, newsletters, advisory cttees, media relations,
188 - mentoring, Perf. Indicators,
189 - Procurement procedure
191 - Social goals are diverse
192 - Conflicts over values - how much support e.g. 60% do 20% - is this adequate?
- Who & how much to gets - disagree
193 market Pkg. - need direction for pkg - what about TP? Transit share declining?
194 - deregulate & restructure transit
195 - summary - new approach
203 - Data Sharing
205 - Regional Imp. Strategy - MPO Income

INDEX

A

AASHO. *See* American Association of State Highway Officials
AASHTO. *See* American Association of State Highway & Transportation Officials.
actual spending, factual data category, 181
Adams, Henry *(The Education of Henry Adams)*, 20n
adaptability, as performance criteria, 189
adaptive discovery, transportation planning process, 169–172
administrative reform movement, 12
advisory committees, 187
Air Board Continues California's World Leadership in Auto Emission Standards, 213n
Air Pollution Control Act, 122
air quality, 4–5
air traffic control system, 14
Alberti *(The Art of Building)*, 18–19
Ambrosiano, Nancy (Lab Gains Corporate Partner for Traffic Simulation), 116n
American Association of State Highway and Transportation Officials (AASHTO), 29n, 50n
American Association of State Highway Officials (AASHO), 34, 44n, 186
American Highway Users Alliance, 49–50, 70n
American Political Science Association (1903), 26
American Renaissance, 21–22
American Road and Transportation Builders Association (ARTBA), 75
American society
  community life, turn of the twentieth century, 19–21
  shift from heavy manufacturing to services, 9–10
American Trucking Associations (ATA), 75
America's Highways, 1776-1976: A History of the Federal-Aid Program, 29n
Ames, Iowa, 1
Analytica software package, 9/n
Anderson, P. W.; Arrow, K. J.; and Pines, D., editors *(The Economy as an Evolving Complex System)*, 142n
anti-statism, 81
Appalachian Mountains, improving access to, 7

Argyris, Chris; Putnam, Robert; and Smith, Diana McLain *(Action Science: Concepts, Methods, and Skills for Research and Intervention)*, 154n
Argyris, Chris (Action Science and Organizational Learning), 154n
Argyris, Chris *(Overcoming Organizational Defenses: Facilitating Organizational Learning)*, 154n
Aristotelian worldview, 17–18
Arrow, Kenneth *(Social Choice and Individual Values)*, 81n
Arrow, Kenneth (The Economic Implications of Learning by Doing), 145n
ARTBA. *See* American Road and Transportation Builders Association
Arthur, W. Brian (Positive Feedbacks in the Economy), 145n
Arthur, W. Brian (Self-Reinforcing Mechanisms in Economics), 142n
Aschauer, David A. (Does Public Capital Crowd Out Private Capital?), 8n
Aschauer, David A. (Is Public Expenditure Productive?), 8n, 137n
Aschauer, David A. (Public Investment and Productivity Growth in the Group of Seven), 8n
Association of Metropolitan Planning Organizations, 224
associative state, 24
ATA. *See* American Trucking Associations
Australia, 188
Austrian School. *See* economics, Austrian School
automatic vehicle identification (AVI), 185
automobiles, demand for, 3
availability, factual data category, 181
AVI. *See* automatic vehicle identification

B

Bad News Principle, 104–106
Bain, David Haward *(Empire Express: Building the First Transcontinental Railroad)*, 7n, 38n
BANANA. *See* build absolutely nothing anywhere near anyone sentiment
Barings Bank, 10

Barnhardt, Cynthia (Workshop on Planning, Design, Management and Control of Transportation Systems), 204n
Barns, Ian (Past-Fordist People? Cultural Meanings of New Technoeconomic Systems), 173
Bartelink, E. H. B., 64n
Basic Financial Statements—and Management's Discussion and Analysis —for State and Local Governments, 225n
Beasley, David R. *(The Suppression of the Automobile: Skullduggery at the Crossroads)*, 146n
beltways, 44–46
benefit-cost analysis, 8, 88
Benjaafar, Saifallah; Morin, Thomas; Talvage, Joseph J. (The Strategic Value of Flexibility in Sequential Decision Making), 173
Benson, Brien, 149n, 179n
Berenson, Bernard, 22
Bernard, L.L. *(Social Control in Its Sociological Aspects)*, 23
Berry, David (The Structure of Electric Utility Least Cost Planning), 100n
Bible, Senator Alan, 57
bioconcentration, 96n
Blum, Justin (Anti-Development Campaigns Take Root), 172n
Boarnet, Marion G. (Highways and Economic Productivity: Interpreting Recent Evidence), 9n, 137n, 211n
Boarnet, Marion G. (Infrastructure Services and the Productivity of Public Capital: The Case of Streets and Highways), 9n, 137n
Boarnet, Marion G. (Road Infrastructure, Economic Productivity, and the Need for Highway Finance Reform), 9n
Boer, Harry and Krabbbendam, Koos (Organizing for Manufacturing Innovation: The Case of Flexible Manufacturing Systems), 173
Boggs, Congressman Hale, 49
Boito, Arrigo (Mephistopheles), 21n
Boston, 26
BOT. *See* build-operate-transfer
Brand, D. (Applying Benefit/Cost Analysis to Identify and Measure the Benefits of Intelligent Transportation Systems), 190n
Brazil, 188

Briggs, Valerie, and Jasper, Keith *(Organizing for Regional Transportation Operations: Arizona AZTech)*, 199n
Briggs, Valerie and Jasper, Keith *(Organizing for Regional Transportation Operations: An Executive Guide)*, 199n
Briggs, Valerie, and Jasper, Keith *(Organizing for Regional Transportation Operations: Houston TranStar)*, 199n
Briggs, Valerie, and Jasper, Keith *(Organizing for Regional Transportation Operations: New York/New Jersey/Connecticut TRANSCOM)*, 199n
Briggs, Valerie, and Jasper, Keith *(Organizing for Regional Transportation Operations: San Francisco Bay Area)*, 199n
Briggs, Valerie, and Jasper, Keith *(Organizing for Regional Transportation Operations: Southern California ITS Priority Corridor)*, 199n
Briggs, Valerie, and Jasper, Keith *(Organizing for Regional Transportation Operations: Vancouver TransLink)*, 199n
Bronx Parkway, 41
Bronzini, Michael (C&P Report Review), 135n
BTS. *See* U.S. Bureau of Transportation Statistics
Buchanan, James M. and Stubblebine, William Craig (Externality), 91n
budget. *See also* financial viability
 factual data category, 181
 highway construction and maintenance, 3
build absolutely nothing anywhere near anyone (BANANA) sentiment, 5
build-operate-transfer (BOT), 99
Bunasekaran, A.; Martikainen, T.; and Yli-Olli, P. (Flexible Manufacturing Systems: An Investigation for Research and Applications), 173
Burchell, Robert W. *(The Costs of Sprawl—Revisited)*, 148
Bureau of Municipal Research (1906), 26
Burnham plan, Chicago, 27

C

CAAA. *See* Clean Air Act Amendments of 1990
CAC. *See* Citizen Advisory Committee
CAD. *See* computer-aided design
Calhoun, Senator John C., 7

California, electric vehicle requirement, 212–213
Canada, bridge between New Brunswick and Prince Edward Island, 186
Canadian Real Return Bonds, 90
canals, providing Appalachian Mountain access, 7
car-friendly communities, 229
carbon monoxide, 86n
career changes, 10
Caro, Robert *(The Power Broker)*, 4n, 13n
Carroll, Kathleen *(Put That Burger on My E-ZPass)*, 202n
CATQEST. *See* Contract Administration Techniques for Quality Enhancement Study Tour
CBO. *See* Congressional Budget Office
CDOH. *See* Colorado Department of Highways
Celebrating America's Highways, 49
Central Artery/Tunnel project, Boston, 185
Central Park carriageways, 41
Chamber of Commerce, 75
chaos
  in American community life, turn-of-the-twentieth-century, 19–21
  Renaissance theories of compared to today's fascination with, 21
Chase Manhattan Bank, 202
Chesapeake and Ohio canal, failure of, 7
Chicago, 1893 World Columbia Exposition, 27
Chicago, Milwaukee, and St. Paul Company, 1
Ciborra, C. U. (Innovation, Networks and Organizational Learning), 201n
CIM. *See* computer integrated manufacturing
circumferential highways. *See* beltways
Citizen Advisory Committee (CAC), 60
Citizens for a Better Environment v. Wilson, 123n
City Beautiful movement, 27, 56
city planning
  American society, 26–27
  Renaissance ideas regarding, 19
Civil Rights movement, 54
civil service reform (1883), 26
Clay, Senator Henry, 7
Clean Air Act, 14
Clean Air Act Amendments of 1970
  Environmental Protection Agency, 122
  federal implementation plans, 122
  importance of, 54, 86n
  state implementation plans, 122
  transportation control plans, 122
Clean Air Act Amendments of 1977

impact on urban transportation planning, 123
  tailpipe emission reductions, 122–123
Clean Air Act Amendments of 1990 (CAAA)
  conformity assessment, 124–126
  evolution of, 71
  transportation control measures, 126–127
  urban transportation planning, integration with air quality planning, 124
CMS. *See* congestion management system
CNC. *See* computerized numerically controlled machines
Coase, Ronald (The Problem of Social Cost), 91n
Coles, Jessie V. *(Standardization of Consumers' Goods: An Aid to Consumer Buying)*, 24n
Colorado Department of Highways (CDOH), 60
Colorado, Glenwood Canyon, 60–63
Columbia Exposition, 27
community needs, viewing as demand side, 2
community separation, 5
computer-aided design (CAD), 165
computer integrated manufacturing (CIM), 165
computerized numerically controlled (CNC) machines, 165
concepts of flexibility, 166
Condition and Performance: 1997 Status of the Nation's Surface Transportation System, 135n, 209n
condition and performance reports (C&P), 134–136, 209, 225
congestion
  costs, 3–4
  factual data category, 181
  performance criteria, 189
  worsening, 3
congestion management system (CMS), 181
Congressional Budget Office (CBO), 66, 208
Connors, Stephen (personal communication), 100n
consortia, 199–201
construction cost, varying, 102–103
consumer price index (CPI), 90
continuity, as performance criteria, 189
Contract Administration Techniques for Quality Enhancement Study Tour (CATQEST), 186n
control, transportation planning process, 164–165, 167–169
Cook, Brian J. *(Bureaucracy and Self-Government: Reconsidering the Role of*

Public Administration in American Politics), 83n
cost reduction, as benefit of publicly funded projects, 8
Cowan, Robin (Tortoises and Hares: Choice among Technologies of Unknown Merit), 146n
Cox, Wendell and Love, Jean *(The Best Investment a Nation Ever Made: A Tribute to the Dwight D. Eisenhower System of Interstate and Defense Highways)*, 50n
Cronon, William, editor *(Uncommon Ground: Toward Reinventing Nature)*, 21n
C&P. *See* condition and performance reports
CPI. *See* consumer price index
crash programs, 66
Crowe, Thomas (Integration Is Not Synonymous with Flexibility), 173
customers, ignoring, 6

D

Dahlman, Carl J. (The Problem of Externality), 91n
Dahms, Lawrence, 121n
Darwin, Charles R. *(On the Origin of the Species)*, 21n
data envelopment analysis (DEA), 96
data interchange standards, 224–225
data sharing model, 203–205
David, Paul A. and Greenstein, Shane (The Economics of Compatibility Standards: An Introduction to Recent Research), 147n
David, Paul A. (Clio and the Economics of QWERTY), 150n
DBOT. *See* design-build-operate-transfer
DEA. *See* data envelopment analysis
Dearing, Charles L. *(American Highway Policy)*, 38n
decision-making, delegating, 10
decision support, 177, 182–184, 219
DeCorla-Souza, Patrick; Everett, Jerry; and Gardner, Brian (Applying a Least Total Cost Approach to Evaluate ITS Alternatives), 100n
dedicated gas taxes, 5
defense capabilities, relationship to roads to, 31
The Delaney Clause, Food Additives Amendment of 1958, 87n

Delucchi, Mark A. *(The Annualized Social Cost of Motor-Vehicle Use in the United States 1990-1991)*, 5n, 8n, 194n
demand, varying, 103–104
demand management techniques, 14
demand side, viewing community needs as, 2
demosclerosis, 5
Des Moines, Iowa, 1
design and implementation, 177, 184–185
design and procurement, 185–186
design-build-operate-transfer (DBOT), 99
design standards
 capacity, 42
 design speed, 40
 grade-separated intersections, 41–42
 limited access, 41
 rights-of-way, 42–43
Design Standards for the National System of Interstate Highways, 44n
Dillon Rule, 57
Disaggregated Residential Allocation Model (DRAM), 115
disorderly socioeconomic systems, 139–140
DiStefano, Joseph R. and Raimi, Matthew *(Five Years of Progress: 110 Communities Where ISTEA is Making a Difference)*, 74n
distributional equality, 89–91
Dixit, Avinash K. and Pindyck, Robert S. *(Investment under Uncertainty)*, 100n
DOT. *See* U.S. Department of Transportation
McShane, Clay *(Down the Asphalt Path: The Automobile and the American City)*, 146n
Downs, Anthony (How Real Are Transit Gains?), 227n
DRAM. *See* Disaggregated Residential Allocation Model
du Pont, Senator Coleman, 41
Dunn, James A. (Driving Forces: The Automobile, Its Enemies, and the Politics of Mobility), 230n

E

E-ZPass, 201–203
Easterlin, Richard A. (A Vision Become Reality), 162n
Eccles, Henry E. *(Military Concepts and Philosophy)*, 66n, 173, 192n
Economic Effects of Transportation, 195n
economic growth, relationship to infrastructure, 7–9

economic productivity studies, 137–138
economic viability, 194–195
economics, Austrian School
   capital structure, 154–156
   choice under uncertainty, 152–153
   competition as learning, 153–154
   neoclassical model, challenges to, 151–152
economies of scale *versus* economies of scope, 10
Edner, Sheldon M., 121n
efficiency fallacy, 163
egalitarianism, 81
EIS. *See* environmental impact statement; environmental impact study
Eisenhower, President Dwight D., 35, 49
EMPAL. *See* Employment Allocation Model
employment
   significant shifts in, 10–11
Employment Allocation Model (EMPAL), 115
energy consumption
   factual data category, 181
   performance criteria, 189
environment
   public's growing concern for, 54
   as reason for decline in urban transport infrastructure public support, 5
environmental applications, 107
environmental approval process, streamlining, 222–223
environmental elitists, 5
environmental impact statement (EIS), 118
environmental impact study (EIS), 62
environmental stewardship, 18
equity, as performance criteria, 189
Erie Canal
   impact, on product distribution, 7
   public opposition to, 4
European Asphalt Study Tour, 186n
executive authority, crisis in, 26
exogenous goal fallacy, 160–161

F

facility designs, public dissatisfaction with, 6
Farrell, Joseph and Saloner, Garth (Standardization, Compatibility and Innovation), 149n
Fatka, Don (From Norway to Story Country), 2n
Federal Aid Highway Act of 1944, 39–40, 50n
Federal Aid Highway Act of 1956, 44, 50n
Federal Aid Highway Act of 1973, 55
Federal Aid Highway Program, 31, 205
Federal Aid Road Act of 1916, 31
Federal Highway Act of 1921, 31–32
Federal Highway Administration (FHWA), 61, 172, 186, 203, 211
federal implementation plans (FIPs), 122
federal level planning
   incentives based on measurable outcomes, 212–213
   information sharing, facilitating, 214
   time horizon, shortening for project analysis, 211–212
   uncertainty, addressing, 208–211
Federal Regional Development Agency, 57
Federal Regulations Impacting Housing and Land Development, Recommendations for Change, Phase I, 24n
federal support, National Road, 1838, 7
Federal Transportation Administration (FTA), 204
FHWA. *See* Federal Highway Administration
financial viability, 189, 194–195. *See also* budget
FIPs. *See* federal implementation plans
Fishlow, Albert *(American Railroads and the Transformation of the Antebellum Economy)*, 94n
Flachsbart, Peter G. (Long-Term Trends in United States Highway Emissions, Ambient Concentrations, and in-Vehicle Exposure to Carbon Monoxide in Traffic), 130n
flexibility, in transportation planning, 165–169, 231–232
Flexibility in Highway Design, 174
flexible manufacturing systems (FMS), 165
Flood Control Act of 1936, 8
FMS. *See* flexible manufacturing systems
Fogel, Robert W. *(Railroads and American Economic Growth: Essays in Econometric History)*, 94n
Food Additives Amendment of 1958, 87n
Food and Drug Administration, 87n
Ford Model T, 24
France, 188
Frank, James E. *(The Costs of Alternative Development Patterns: A Review of the Literature)*, 148n
Freeway Revolt, 53–55
Friedlaender, Ann F. *(The Interstate Highway System: A Study in Public Investment)*, 94n
Friedman, John (Planning, Progress, and Social Values), 11–12

Fukuyama, Francis, 13, 200n
Fuller, Stephen, 11n
fund accounting, 12

G

Gaillard, John *(Industrial Standardization: Its Principles and Applications)*, 25n
Garratt, Bob *(Creating a Learning Organization: A Guide to leadership, Learning and Development)*, 154n
Garrett, Mark and Wachs, Martin *(Transportation Planning on Trial: The Clean Air Act and Travel Forecasting)*, 123n
gas taxes, 5
GASB. *See* Government Accounting Standards Board
gasoline tax, 32
Geddes, Patrick, 26n
general fund support, 5
General Location of National System of Interstate Highways, Including All Additional Routes at Urban Areas, 46n
geographic information systems (GIS), 204, 220
Georgia Department of Natural Resources, 204
Gertler, Meric S. (The Limits to Flexibility: Comments on the Post-Fordist Vision of Production and Its Geography), 167n, 173
Gerwin, Donald (An Agenda for Research on the Flexibility of Manufacturing Processes), 173
Gifford, J. J.; Yermack, L., and Owens, C. (The Development of the E-ZPass Specification in New York, New Jersey and Pennsylvania: A Case Study of Institutional and Organizational Issues), 202n
Gifford, Jonathan L. (An Analysis of the Federal Role in the Planning, Design and Deployment of Rural Roads, Toll Roads and Urban Freeways), 30n, 39n
Gifford, Jonathan L. and Pelletiere, Danila ("New" Regional Transportation Organizations: Old Problem, New Wrinkle), 205n
Gifford, Jonathan L. and Pelletiere, Danila (Study on Innovative Policies in the U.S.), 126n
Gifford, Jonathan L. and Stalebrink, Odd J. (Remaking Transportation Organizations for the 21st Century: Learning Organizations and the Value of Consortia), 154n, 199n
Gifford, Jonathan L. (Complexity, Adaptability, and Flexibility in Infrastructure and Regional Development: Insights and Implications for Policy Analysis and Planning), 140n
Gifford, Jonathan L.; Hardy, Matthew; and Owens, C. (Voluntary Organizations in the Deployment of Intelligent Transportation Systems), 206n
Gifford, Jonathan L.; Horan, Thomas A.; and White, Louise G. (Dynamics of Policy Change: Reflections on 1991 Federal Transportation Legislation), 69n
Gifford, Jonathan L.; Horan, Thomas A. (Transportation and the Environment), 80n
Gifford, Jonathan L.; Mallett, William J.; and Talkington, Scott W. (Implementing Intermodal Surface Transportation Act of 1991: Issues and Early Fields Data), 74n
Gifford, Jonathan L. (The Innovation of the Interstate Highway System), 30n
Gilbreth, Frank, 23
Gilbreth, L.E.M *(The Psychology of Management: The Function of the Mind in Determining, Teaching and Installing Methods of Least Waste)*, 23n
GIS. *See* geographic information systems; graphic information systems
Glenwood Canyon, Colorado (35-mm film), 61n
goals and means, framework of analyzing, 168
going cold turkey, 198
Golden Age, 51–53
Goodnow, Frank J. *(Politics and Administration: A Study in Government)*, 25n
Gordon, John Steele (Through Darkest America), 35n
Gore, Senator Albert Sr., 49
Governance in Transition: Public Management Reforms in OECD Countries, 12n, 188n
government, changing roles and responsibilities, 12
Governmental Accounting Standards Board (GASB), 225
Government Performance and Results Act of 1993 (GPRA), 12, 188

GPRA. *See* Government Performance and Results Act of 1993
Gramlich, Edward M. (Benefit-Cost Analysis of Government Programs), 88n
Gramlich, Edward M. (Infrastructure Investment: A Review Essay), 8n
graphic information systems (GIS), 220
Great Britain, nineteenth-century toll roads, 38
Great Depression, 33
Great Society Programs, 26
Gribeauval, Jean-Baptiste de, 24
Guam, 22
Guidance for Communicating the Economic Impacts of Transportation Investments, 187n

H

Hadden, Samuel, C., 34n
Haber, Samuel *(Efficiency and Uplift),* 23
Harriman, Norman Follett *(Standards and Standardization),* 23n
Hartgen, David T. and Lindeman, Nicholas J. *(The ISTEA Legacy: Comparative Performance of State Highway Systems, 1984-1997),* 189n
Harvey, Greg and Deakin, Elizabeth (Air Quality and Transportation Planning: An Assessment of Recent Developments), 123n
Haughton, G. and Browett, J. (Flexible Theory and Flexible Regulation: Collaboration and Competition in the McLaren Vale Wine Industry in South Australia), 173
Haussmann, Baron, 4, 45
Hawley, Ellis W. (Herbert Hoover, the Commerce Secretariat and the Vision of an "Associative State"), 24n
Hayek, Friedrich A. von (Competition as a Discovery Procedure), 153n
Hayek, Friedrich A. von (The Use of Knowledge in Society), 152n
Haynes, Kingsley E. (Planning for Capacity Expansion: Stochastic Process and Game Theoretic Approaches), 96n, 185n
Hays, Samuel P. *(Conservation and the Gospel of Efficiency),* 22n
Heanue, Kevin (conversation), 71n, 75n
Heanue, Kevin *(Data Sharing and Data Partnerships for Highways),* 203n
Heanue, Kevin (A Review of the Conditions & Performance Reporting Process), 135n

Hebden, Norman and Smith, Wilber S. *(State-City Relationships in Highway Affairs),* 47n
HERS. *See* Highway Economic Requirement System
High Needs of the National Defense, 44n
High Occupancy Tolls (HOT-Lanes) and Value Pricing: A Preliminary Assessment, 91n
high occupancy vehicle (HOV), 59, 179
Highland Park factory, Detroit, 24
Highway and Transit Environmental Streamlining Progress Summary, 172n, 223n
highway capital stocks, deficient data model, 9
highway construction and maintenance budgets, 3
Highway Economic Requirement System (HERS), 225
highway needs studies, 131–134
Highway Performance Monitoring System (HPMS), 134, 203, 225
Highway Performance Monitoring System Reassessment, 214n
highway planning, before World War II, 111–112
Highway Revenue Act of 1956, 50n
Highway Statistics 1994, 58n
Highway Trust Fund (1956), 44, 46, 50–51, 210
Hilts, H.E. (Planning the Interregional Highway System), 40n
The History and Accomplishment of Twenty-Five Years of Federal Aid for Highways: An Examination of Policies from State and National Viewpoints, 34n
Holland, John (Can There Be a Unified Theory of Complex Adaptive Systems?"), 140n
honesty, 177–180
Hoover, Herbert, 24
Hott, Larry and Lewis, Tom *(Divided Highways),* 50n
Hounshell, David A. *(From the American System to Mass Production, 1900-1932: The Development of Manufacturing Technology in the United States),* 24n
Hounshell, David A. (personal communication), 94n
household structure, shifting, 11
HOV. *See* high occupancy vehicle
Howitt, Arnold M. and Altshuler, Alan (Controlling Auto Air Pollution), 118n

239

Howitt, Arnold M. and Moore, Elizabeth M. (The Conformity Assessment Project: Selected Findings), 128n
HPMS. *See* High Performance Monitoring System
Hughes, Thomas P. *(American Genesis: A Century of Invention and Technological Enthusiasm)*, 23
hydrocarbons, 86n

I

IAG. *See* Interagency Group
IBTTA. *See* International Bridge, Tunnel and Turnpike Association
income elastic, 5
increasing returns
 adaptive expectations, 147
 coordination effects, 142–144
 economies of scale and scope, 144–145
 importance of, 140–142
 learning by doing, 145–146
 lock in, 149–150
 multiple equilibria, 147–148
 path dependence, 150–151
 possible inefficiency, 148
 sources of, 142
indicative planning, 85–86
indicators, capturing, 218–219
individualism, 81
industrial organization, self-organizational approaches, 10
Industrial Revolution, impact upon small town life, 20
informal voluntary organizations, 206–207
information quality, 93–94
infrastructure planning, neoclassical model, 156–157
infrastructure suppliers, failure to deliver publicly-acceptable facilities, 6
Innovations in Delivery of Transportation Services, 189n
Innovative Contracting Practices, 186n
Integrated Land Use Transportation Package (ITLUP), 115
integrated resource planning (IRP), 99
intelligence, as source of factual information, 180–182
intelligence function, 218
Intelligent Transportation: Realizing the Future, 201n
intelligent transportation system (ITS), 94, 172, 185, 199, 223

Interagency Group (IAG), 201–203
Intermodal Surface Transportation Efficiency Act of 1991 (ISTEA)
 changes brought about by, 73–74
 congestion management system, 181
 local officials, strengthening role of, 31
 metropolitan planning organizations, increasing role of, 120–121
 metropolitan transportation planning, 119
 Nickel for America, 71
 passage of, 73
 pork barrel projects, 69
 public involvement, 120
 Senate Committee on Environment and Public Works, 72–73
 spectrum of planning practice, 120
 Surface Transportation Policy Project, 72
 Transportation 2020, 70
International Bridge, Tunnel and Turnpike Association (IBTTA), 203
Internet, impact upon data availability, 207
Interregional Highways, 39
Interstate 66, 56–60, 186
Interstate 70, 55, 60–63
Interstate 93, 55, 63–65
Interstate Commerce Commission (1887), 26
Interstate highway system
 beltways, 44–46
 design standards, 40–43
 finishing, 65
 freeways, backlash against, 6
 funding, 44
 institutional arrangements, 46–47
 as last great public works project, 8
 limited mileage, 43–44
 presidential administrations, role of, 35
 return on investment, 8–9
 toll roads *versus* free roads, 37–38
 urban travel modeling, impact of, 117–118
 urban *versus* interregional highways, 39–40
The Interstate Highway System: Issues and Options, 67n
investing under uncertainty, 100–102, 107–108
IRP. *See* integrated resource planning
irreversibility, 95–96
ISTEA. *See* Intermodal Surface Transportation Efficiency Act of 1991
Italian Renaissance, order and efficiency, 17–19
ITLUP. *See* Integrated Land Use Transportation Package
ITS. *See* intelligent transportation system

J

Jacoby, Arthur, 9n
James, Jeffrey and Bhalla, Ajit (Flexible Specialization, New Technologies and Future Industrialization in Developing Countries), 173
Jefferson, Thomas, 24
Johansson, Börje, 164n
Johnson, Hildegard Binder *(Order upon the Land: The U.S. Rectangular Land Survey and the Upper Mississippi Country),* 150n
just-in-time inventory management, 8

K

Kaboolian, Linda (The New Public Management: Challenging the Boundaries of the Management vs. Administration Debate), 12n, 188n
Kain, John F. (Deception in Dallas: Strategic Misrepresentation in Rail Transit Promotion and Evaluation), 119n, 178n
Kapur, S. (Of Flexibility and Information), 173
Kelman, Steven *(Procurement and Public Management),* 12n, 189n
Kelman, Steven (White House-Initiated Management Reform: Implementing Federal Procurement Reform, 12n, 189n
Kemp, Martin (The Mean and Measure of All Things), 18n
Kirzner, Israel M. *(Austrian School of Economics),* 12n
Kirzner, Israel M. *(Competition and Entrepreneurship),* 12n
Klein, B. H. *(Dynamic Economics),* 201n
Klein, Daniel B.; Moore, Adrian; and Reja, Binyam *(Curb Rights: A Foundation for Free Enterprise in Urban Transit)* 180n, 194n
Knodel, Walter *(Graphentheoretische Methoden un Ihre Anwendungen),* 141n
Konvitz, Josef W. *(The Urban Millennium: The City-Building Process from the Early Middle Ages to the Present),* 7n, 149n
Korea, 188
Kriz, Margaret (Road Warriors), 76n
Krugman, Paul (Increasing Returns and Economic Geography), 148n
Krutilla, John V. and Fisher, Anthony C. *(The Economics of Natural Environments: Studies in the Valuation of Commodity and Amenity Resources),* 93n
Kula, Erhun (Discount Factors for Public Sector Investment Projects Using the Sum of Discounted Consumption Flows: Estimate for the United Kingdom), 89n
Kuznets, Simon *(Modern Economic Growth: Rate, Structure and Spread),* 162n

L

La Farge, John, 22
Lachmann, Ludwig M. *(Capital and Its Structure),* 154n
Lakshmanan, T. R. and Anderson, William P. (Transpiration Infrastructure, Freight Services Sector and Economic Growth), 195n, 211n
land grants, federal support for railroad expansion, 7
land use problems, 5
lanes, number of, 42
large infrastructure projects, public opposition to, 4
Lathrop, William H. (The San Francisco Freeway Revolt), 54n
Lavoie, Don (conversation), 152n, 154n
Lavoie, Don *(National Economic Planning: What is Left?),* 153n
LCCA. *See* life-cycle cost analysis
least cost planning (LCP), 99–100
LeBlanc, L. J. (An Algorithm for Discrete Network Design), 141n
Lee, Douglass B. Jr. (A Requiem for Large-Scale Models), 130n
Lemer, Andrew C. (Building Public Works Infrastructure Management Systems for Achieving High Return on Public Assets), 12n
Lemer, Andrew C. (Major Components of the Infrastructure Asset-Management Process), 189n
L'Enfant Plan of 1971, 56
level of service (LOS), 133
Lewin, R. (Complexity: Life at the Edge of Chaos), 140n
Lewis, Tom *(Divided Highways: Building the Interstate Highways, Transforming American Life),* 50n
Liebowitz, S. J. and Margolis, Stephen E. (Are Network Externalities a New Source of Market Failure?), 151n

Liebowitz, S. J. and Margolis, Stephen E. (Path Dependence, Lock-in and History), 151n
Liebowitz, S. J. and Margolis, Stephen E. (The Fable of the Keys), 151n
life-cycle cost analysis (LCCA), 98–99, 223
Lim, S. H. (Flexible Manufacturing Systems and Manufacturing Flexibility in the United Kingdom), 173
Lindberg, Per (Management of Uncertainty in AMT Implementation: The Case of FMS), 173
linguistic imprecision, 92
Lipset, Seymour Martin (American Exceptionalism Reaffirmed), 81n
loans, federal support for railroad expansion, 7
Lockheed Martin IMS, 202
Lockwood, Stephen C. *(The Changing State DOT)*, 186n
Lomax, Tim *(Quantifying Congestion)*, 181n
LOS. *See* level of service
Luchian, Serge; Krechmer, Daniel; and Muzzey, Paul (Case Study of Electronic Toll Collection in the Central Artery/Tunnel Project—Boston), 185n
Lukacs, John (The Gotthard Walk), 21n
Lumina Decision Systems, 97n
Lyons, William M. (The FTA-FHWA MPO Reviews: Planning Practice under the Intermodal Surface Transportation Efficiency Act and the Clean Air Act Amendments), 121n

M

MacDonald, Thomas H., 38
MacDonald, Thomas H. (Flatten Out Those Traffic Peaks: Highways of Tomorrow Will Avoid Big City Congestion and Provide Safe Speed in the Country), 40n
MacDonald, Thomas H. (letter to Colonel James Roosevelt), 36n
MacDonald, Thomas H. (The Freedom of the Road), 37n
Mack, W. W., 41
MacRae, Duncan Jr. and Wilde, James A. (Perfect Markets, Imperfect Markets, and Policy Corrections), 37n
macroeconomic analysis, 225
Mahler, Julianne (Influences of Organizational Culture on Learning in Public Agencies), 154n
mail, free rural delivery, 1896, 20

Malecki, Edward J. (The Location of Economic Activities: Flexibility and Agglomeration), 173
management by results, 10
Mandelbaum, Marvin and Buzacott, John (Flexibility and Decision Making), 173
margin of safety analysis, 87–88
Mark IV tags, 202
Market-Based Approach to Zero Emission Vehicle Program Working for Californians, 213
Marquardt, Michael *(Building the Learning Organization: A Systems Approach to Quantum Improvement and Global Success)*, 154n
Maryland–National Capital Park and Planning Commission, 56
McDonald's, accepting E-ZPass as form of payment, 202
McMillan plan, for Washington, D.C., 27
McNally, Michael G. (The Activity-Based Approach), 160n
McShane, Clay *(Down the Asphalt Path: The Automobile and the American City)*, 29–30n
McShane, Clay (The Failure of the Steam Automobile), 146n
media relations, 187
MEPLAN, 115
Merkhofer, M. W. (The Value of Information Given Decision Flexibility), 174
Metcalf, Gilbert E. and Rosenthal, Donald (The "New" View of Investment Decisions and Public Policy Analysis: An Application to Green Lights and Cold Refrigerators), 100n
metropolitan planning organization (MPO), 57, 74, 205–207, 220
metropolitan planning process, deregulating, 222
metropolitan transportation planning, 112–116
Metropolitan Transportation Planning under ISTEA: The Shape of Things to Come, 121n
METROSIM, 115
MFS Network Technologies, 202
mileage, limited, 43–44
military intervention, Persian Gulf, 5
Miller, John B. *(Principles of Public and Private Infrastructure Delivery)*, 186n
Mississippi River, steamboat service, 7
mobile sites, impact on employment, 11

mobile society, compared to immobile society, 4
monitoring, 177, 188, 190–194, 221
Morgan, M. Granger and Henrion, Max *(Uncertainty: A Guide to Dealing with Uncertainty in Quantitative Risk and Policy Analysis)*, 91n
Morison, S.E. and Commager, H.S. *(The Growth of the American Republic*, 4th ed.), 8n
Morison, S.E.; Commager, H.S.; and Leuchtenburg, W.E. *(The Growth of the American Republic*, 7th ed.), 12n
Morowitz, Harold *(Entropy and the Magic Flute)*, 18n
Moses, Robert, 4, 13
Moving America: New Directions, New Opportunities, 71n
Moynihan, Daniel P., 72
MPO. *See* metropolitan planning organization
Mudge, Richard R. and Griffin, Cynthia S. (Approaches to the Economic Evaluation of IVHS Technology), 89n
Mudge, Richard R. (Assessing Transport's Economic Impact: Approaches and Key Issues), 8n
Mueller, Norman C. (Fifteen Years of HPMS Partnership: Accomplishments and Future Directions), 203n
Multimodal Transportation Development of a Performance-Based Planning Process, 171n
Multimodal Transportation Planning Data, 204
Mumford, Lewis *(Technics and Civilization)*, 24n
Mumford, Lewis *(The City in History)*, 18n
Murchland, J. D. (Braess' Paradox of Traffic Flow), 141n
Murdoch, J. (Actor-Networks and the Evolution of Economic Forms: Combining Description and Explanation in Theories of Regulation, Flexible Specialization, and Networks), 174
Musil, Robert, 69

N

NAAQS. *See* National Ambient Air Quality Standards
Nadiri, M. Ishaq and Mamuneas, Theofanis P. *(Contribution of Highway Capital Infrastructure to Industry and National Productivity Growth)*, 9n

Nadiri, M. Ishaq and Mamuneas, Theofanis P. (Highway Capital and Productivity Growth), 9n
Nagel, Jack H. (The New Public Management in New Zealand and Beyond), 188n
Napoleon III, 4
National Ambient Air Quality Standards (NAAQS), 86, 122, 212
National Association of Manufacturers, 75
National Association of Regional Councils, 224
national campaign, 198–199
National Capital Metropolitan Conference, 57
National Capital Planning Commission (NCPC), 56
National Capital Regional Planning Council (NCRPC), 56
National Capital Transportation Agency (NCTA), 57
National Cooperative Highway Research Program, 171n
National Dialog on Transportation Operations (2001), 172
National Environmental Policy Act of 1969 (NEPA), 54, 64, 118
National Functional System Mileage and Travel Summary: From the 1976 National Highway Inventory and Performance Study, 132n
National Governors Association, 224
National Highway System Designation Act of 1995, 74–75, 181n
National Highway System (NHS), 74
National Institute of Building Sciences, 24n
National Interregional Highway Committee, 39
National League of Cities, 224
national-level recommendations, 221–226
National Library of Medicine, 205
National Municipal League (1984), 26
National Performance Review, 12
National Road, Maryland to Ohio, 7
National Summit on Transportation Operations, 172
nationalism, during American Renaissance, 22
NCPC. *See* National Capital Planning Commission
NCRPC. *See* National Capital Regional Planning Council
NCTA. *See* National Capital Transportation Agency
Nelson, Dick and Shakow, Don (Least-Cost Planning: a Tool for Metropolitan Transportation Decision Making), 100n

NEPA. *See* National Environmental Policy Act of 1969
Neudorff, Louis G. and Batz, Tom (A Regional ITS Architecture for the New York Metropolitan Area), 201n
Neumann, Francis X. Jr. (Organizational Responses to Complex and Changing Environments in the Great Lakes Basin), 199–200n
New Brunswick, bridge to Prince Edward Island, 186
New Hampshire, Franconia Notch, 63–65
New Hampshire State Department of Transportation, 64n
New Transportation Concepts for a New Century: AASHTO Recommendations on the Direction of the Future Federal Surface Transportation Program and for a National Transportation Policy, 70n
new urbanists, 229
New York City, 4, 7
Newman, Peter W. G. and Kenworthy, Jeffrey R. *(Cities and Automobile Dependence: A Sourcebook)*, 148
Newman, W. Rocky; Hana, Mark; and Maffei, Mary Jo (Dealing with the Uncertainties of Manufacturing: Flexibility, Buffers and Integration), 174
newsletters, 187
NHS. *See* National Highway System
Nickel for America, 71
NIMBY. *See* not in my backyard sentiment
nitrogen dioxide, 86n
noise, 4, 8
Noise Control Act of 1972, 54
non-pecuniary impacts, 95–96
non-scale planning, 130–131
non-subsistence amenities, value places upon, 5
Northern Virginia Regional Plan: Year 2000, 56
Northern Virginia Regional Planning and Economic Development Commission, 56
Northwestern Company, 1
not in my backyard (NIMBY) sentiment, 5

O

Oakley, Janet, 71n, 72n
object lesson road, 30
OECD. *See* Organization for Economic Cooperation and Development

Ohio River, steamboat service, 7
oil crisis, 54–55
O&M. *See* operations and management
OMB. *See* U.S. Office of Management and Budget
On Wedges and Corridors: A General Plan for the Maryland—Washington Regional District, 56n
operations and management (O&M), 171
order and efficiency, Italian Renaissance theories regarding, 17–19
Organization for Economic Cooperation and Development (OECD), 12
ornamentation, 27
Orski, Ken (Developing an "Operations Vision": The Dialog Continues), 172n
Oryani, Kazem and Harris, Britton *(Review of Land Use Models and Recommended Model for DVRPC)*, 115n
Ott, W. and Roberts, John W. (Everyday Exposure to Toxic Pollutants), 129n
Ott, W. and Willits, N. (Carbon-Monoxide Exposures inside an Automobile Traveling on an Urban Arterial Highway), 129n
outcomes, measurable, 222
outreach and community involvement, 186–187
ozone, 86n

P

Panama Canal, 22
Pansing, C.; Schreffler, E. N. P.; and Sillings, M. A. (Comparative Evaluation of the Cost-Effectiveness of 58 Transportation Control Measures), 126n
Paris, 4
participatory decision making, 108–110
pavement condition, factual data category, 181
peak transportation hours, expanding to longer periods, 3
Pennsylvania, Philadelphia, 7
Pennsylvania, Pittsburg, 7
Pennsylvania Turnpike, 37–38
people-friendly communities, 229
Pérez-Peña, Richard (Pataki Seeks U.S. Waiver on Air Quality), 170n
performance, emphasis upon, 12
performance-based evaluation, 10
performance indicators, 188–190
Perlman, Ellen (Breakdown in the ZEV Lane), 213n
Pershing, General John, 35

Philippines, 22
plans, specifications and estimates (PS&E), 79
Platonic worldview, 17–18
A Policies Plan for the Year 2000, 56
policing service, 5
populism, 81
pork barrel projects, 69
Port Authority of New York and New Jersey, 202
Porter, Alan F. et al. *(Effects of Federal Transportation Funding Policies and Structures: Final Report),* 117n
power, impact upon transportation infrastructure construction, 4
prediction-based analytical planning, 14
predictive modeling fallacy, 161–163
Pressman, Jeffrey L. and Wildavsky, Aaron *(Implementation: How Great Expectations in Washington are Dashed in Oakland),* 199
Prince Edward Island, bridge to New Brunswick, 186
principal-agent theory, 12
probabilistic risk assessment, 96–98
procedural integrity, emphasizing over results and performance, 12
product cycle, reducing, 10
production efficiency, 24
productivity measurement system, 9
Progressive Era
  ideology, 22
  public administration, 25–26
  scientific management, 23
  standardization, 23–25
project cost–benefit analysis, shortening time for, 223
public
  consensus regarding urban transportation infrastructure, 3
  opposition large infrastructure projects, 4
  trust in government, 5
  unwillingness to confront traffic growth causes, 5
  unwillingness to defer to government experts, 13
public administration, 25–26
public choice theory, 12
public involvement fallacy, 163–164
Public Management Developments: Update 1994, 12n, 188n
public sector reforms, worldwide, 12
public support, reasons for erosion of, 4–5
public transit
  erosion of viable markets, 5
  public support for, 6
Puerto Rico, 22
Putnam, Robert D. *(Bowling Alone: The Collapse and Revival of American Community),* 81n
Pye, Roger (A Formal, Decision-Theoretic Approach to Flexibility and Robustness), 174

Q

Quadrennial Defense Review (1997), 165
quality of life, 2
Quinn, James Brian (Strategic Change: "Logical Incrementalism"), 174, 192n

R

railroad expansion, early support for, 7
RAP. *See* Risk Analysis Package
Rauch, Jonathan *(Demosclerosis: The Silent Killer of American Government),* 5
raw data, summarizing, 219
Rawls, John *(A Theory of Justice),* 89n
Reagan, Patrick D. *(Designing a New America: The Origins of New Deal Planning, 1890-1943),* 112n
Reagan, President Ronald, 66
Realistic Goals: A Revised Electric-Car Mandate, 213n
Reconstruction Finance Corporation (RFC), 37
regional conferences, 218
regional data, making available, 219
The Regional Development Guide 1966-2000, 56
Reinventing Government, 12
reliability, as performance criteria, 189
Reno, Arlee (Potential Improvements—C&P/Reauthorization Bottom Line), 135n
Report of the Quadrennial Defense Review, 165n
Report on the 1992 U.S. Tour of European Concrete Highways—U.S. TECH, 186n
Reps, John William *(The Making of Urban America: A History of City Planning in the United States),* 26n
RFC. *See* Reconstruction Finance Corporation
Risk Analysis Package (RAP), 96n
Roosevelt, Colonel James, 36n
Roosevelt, President Franklin D., 35
Roosevelt, Teddy, 22

Rose, Mark H. *(Interstate Express Highway Politics, 1941-1989)*, 29n, 44n, 50n
Rosen, Christine Meisner *(The Limits of Power: Great Fires and the Process of City Growth in America)*, 149n
Rosenburg, Nathan *(Inside the Black Box: Technology and Economics)*, 145n
roundabouts, 42
Rubin, Paul *(Managing Business Transactions: Controlling the Cost of Coordinating, Communicating and Decision Making)*, 185n
Russell, Jeffrey Burton *(Mephistopheles: The Devil in the Modern World)*, 21n

S

Saalman, Howard *(Haussmann: Paris Transformed)*, 4n, 45n
safety
  factual data category, 181
  performance criteria, 189
scale *versus* flexibility, 106–107
Schieve, W. C. and Allen, P. M. *(Self-Organization and Dissipative Structures: Applications in the Physical and Social Sciences)*, 140n
Scherer, F. M., 211n
Scott, Mel *(American City Planning since 1890)*, 26n
Secular Trinity, 34–35
Seely, Bruce E. *(Building the American Highway System: Engineers as Policy Makers)*, 24n, 29n
self-employment, 11
self-organizational approaches
  government, 12–13
  manufacturing industry, 10
sen, John et al. (Improved Method for Measuring Highway Agency Performance), 189n
Senge, Peter M. *(The Fifth Discipline: The Art and Practice of the Learning Organization)*, 154n
Senge, Peter M. (The Leader's New Work: Building Learning Organizations), 154n
service industry, compared to heavy manufacturing industry, 9–10
Shafritz, Jay M. and Hyde, Albert C. *(Classics of Public Administration)*, 25n
Sheldahl, Osmund, 1

Sheriff, Carol *(The Artificial River: The Erie Canal and the Paradox of Progress, 1817-1862)*, 4n
Sherman, Leonard, 117n
Shumate, Charles, 60n
Sierra Club v. Metropolitan Transportation Commission, 123n
signalized intersections, 42
single occupant vehicle (SOV), 14
SIPs. *See* state implementation plans
Skorstad, Egil (Mass Production, Flexible Specialization and Just-in-Time: Future Development Trends of Industrial Production and Consequences on Conditions of Work), 174
slums, 43
Smith, Wilbur, 46–47
social impact, as reason for decline in support for urban transport infrastructure, 5
societies, high-trust, 200
SOV. *See* single occupant vehicle
Spanish–American War, 22
Sparrow, Malcolm *(The State/EPA Data Management Program)*, 182n, 205n
speed
  factual data category, 181
  performance criteria, 189
stability and agility, 176–177
stakeholders, inventory, 218
Stalebrink, Odd J. and Gifford, Jonathan L. (Actors and Directions in U.S. Transportation Asset Management), 172n
standardization, Progressive Era, 23–25
State and Metropolitan Area Data Book, 1997-98, 201n
state implementation plans (SIPs), 122
states, role of, 207–208
The States and the Interstates: Research on the Planning, Design and Construction of the Interstate and Defense Highway System, 29n, 50n, 70n
states rights, 7
steamboat service, 7
Stevenson, Robert Lewis (The Day After Tomorrow), 192n
Stillman, Richard J. *(Public Administration*, 4th ed.), 25n
Stone, Peter H. (From the K Street Corridor), 75n
Stout, Russell (Management, Control and Decision), 168

STPP. *See* Surface Transportation Policy Project
street cleaning service, 5
subsidies, federal support for railroad expansion, 7
SULEV. *See* super ultra low emission vehicle
sulfur dioxide, 86n
super ultra low emission vehicle (SULEV), 213
supply side, viewing transportation infrastructure as, 2
Surface Transportation Policy Project (STPP), 72
Sweden, 188
Szoboszlay, Akos (HOV Lanes Cause Huge Solo Driver Increase on Montague Expressway), 126n

T

tailpipe emissions, 8
tariff remission on rails, 7
Taylor, Frederick *(Principles of Scientific Management)*, 23n
TCM. *See* transportation control measures
TCPs. *See* transportation control plans
TEA-21. *See* Transportation Equity Act for the 21st Century
TEA 21 Restoration Act, 31n
technical review group (TRG), 62
telecommuting, 11
temporary workers, 10
TERMs. *See* Transportation Emissions Reduction Measures
Thünen, Johann Heinrich von *(The Isolated State),* 141n
time of day, impact upon travel volumes, 3
time savings, as benefit of publicly funded projects, 8
TIPS. *See* Treasury Inflation Protection Securities
TMIP. *See* Travel Model Improvement Program
Tocqueville, Alexis de *(Democracy in America),* 81n
Toll Roads and Free Roads: Message from the President of the United States, 34n, 36
toll roads *versus* free roads, 38–39
Tomlinson, Jim, 112n
Tonn, Bruce E. (500-Year Planning: A Speculative Provocation), 89n
total suspended particulates (TSPs), 86n
town planning, during Middle Ages, 18
traffic circles, 42

traffic counts
　factual data category, 181
　performance criteria, 181
traffic planning, competing roles of, 183
transaction cost economic theory, 12
TRANSCOM (Transportation Operations Coordinating Committee), 201
TRANSIMS, 116
Transportation 2020, 70
Transportation and National Policy, 131n
transportation control measures (TCMs), 14, 124
transportation control plans (TCPs), 122
transportation demand, excess capacity, 3
Transportation Emissions Reduction Measures (TERMs), 126
Transportation Equity Act for the 21st Century (TEA-21), 31, 68–69, 75–76, 223. *See also* TEA 21 Restoration Act
transportation funding, increases, 6
transportation infrastructure, as determinant of economic and social well-being, 2
Transportation Law Benefits Those Who Held the Purse Strings, 76n
transportation planning
　air quality, controlling, 128–130
　analysis, role of, 83–85
　infrastructure and economic growth, 7–9
　institutional context of, 80–83
Transportation Research Board (TRB), 186
Traub, J. F. (Complexity of Approximately Solved Problems), 152n
travel, reasons for undertaking, 4–5
Travel Model Improvement Program (TMIP), 116, 183
TRB. *See* Transportation Research Board
Treasury Inflation Protection Securities (TIPS), 90
TRG. *See* technical review group
truck travel, 67–68
Truman, President Harry S., 35
TSPs. *See* total suspended particulates
Turner, Frank, 49

U

Uchitelle, Louis (Temporary Workers Are on the Increase in Nation's Factories), 11n
uncertainty,
　increasing over demand, 104
　sources of, 91–95

Unter, Brian (The Importance of Standards to Hewlett-Packard's Competitive Business Strategy), 200n
Urban Mass Transportation Administration, 119n
Urban Rail Transit Projects: Forecast versus Actual Ridership and Costs, 119n, 178n
urban transportation dilemma, traffic demand *versus* highway supply, 2
urban transportation planning
  need for new approach, 13–15
  reinventing, 14
Urban Transportation Planning in the United States: An Historical Overview, 30n
urban *versus* interregional highways, 39–40
U.S. Bureau of Census, 201n
U.S. Bureau of Economic Analysis, 3
U.S. Bureau of Public Roads
  creation of, 33
  late-Depression era, 34–36
  toll roads *versus* free roads, 37–38
U.S. Bureau of Transportation Statistics (BTS), 214, 225
U.S. Congress
  Congressional Budget Office, 66, 67n, 208
  increases in transportation funding, 6
  policy formulation, 625
  Roads, hearings on H.R. 7079, 36n
  U.S. House Committee on Public Works, 44n
  U.S. House Committee on Roads, 35–36, 39n
U.S. Department of Transportation (DOT), 30n, 61, 71n, 109n, 209–211
U.S. Department of Treasury, 208
U.S. Disbursements for Highways as a Percentage of GDP: 1945-1998, 3
U.S. Environmental Protection Agency (EPA), 122, 204–205
U.S. Federal Highway Administration, 9, 29n, 58n
U.S. Federal Highway Administration and Federal Transit Administration, 109n, 135n
U.S. Federal Works Agency, 45n, 46n
U.S. National Resources Planning Board, 131
U.S. Office of Management and Budget (OMB), 75, 208
U.S. Office of Road Inquiry, 30
USA Air Quality Nonattainment Areas, 125n

V

Van Riper, Paul *(The American Administrative State: Wilson and the Founders—An Unorthodox View),* 25n
Veblen, Thorsten *(The Engineers and the Price System),* 23
vehicle trips and travel *versus* trip length, 67
Vietnam War, 26, 55
volunteer organizations, 199–201

W

Wachs, Martin (Ethics and Advocacy in Forecasting for Public Policy), 119n, 178n
Wachs, Martin (When Planners Lie with Numbers), 119n, 178n
Waldrop, M. M. (Complexity: The Emerging Science at the Edge of Order and Chaos), 140n
Wallace, Henry, 39n
Walls, James III and Smith, Michael R. *(Life-Cycle Cost Analysis in Pavement Design: Interim Technical Bulletin),* 99n, 223n
Walras, Léon *(Elements of Pure Economics: or, The Theory of Social Wealth),* 21n
Walton, C.M. *(Emerging Models for Delivering Transportation Programs and Services: A Report of the Transportation Agency Organization and Management Scan Tour),* 12n
Washington, D.C.
  Interstate 66, 56–60
  McMillan plan, 27
  self-employment statistics, 11
Watergate scandal, 26, 55
Weimer, David Leo and Vining, Aidan R. *(Policy Analysis: Concepts and Practice),* 82n, 88n
Weiner, Edward *(Urban Transportation Planning in the United States: An Historical Overview),* 30n, 112n
Weingroff, Richard F. (Broader Ribbons across the Land), 35n
Weingroff, Richard F. (Three States Claim First Interstate Highway), 51n
Wellington, Arthur M. *(The Economic Theory of Location of Railways: An Analysis of the Conditions Controlling the Layout Out of Railways to Effect the Most Judicious Expenditure of Capital),* 136n

White, Louise, 167n
White, Louse G. (Policy Analysis as Discourse), 85
White, Stanford, 22
Whitman et al. v. American Trucking Associations, Inc. et al., 125n
Whitney, Eli, 24
Wiebe, Robert H. *(The Search for Order: 1877-1920)*, 19n
Wildavsky, Ben (Pigging Out), 75n
Wiley, Marion C. *(The High Road)*, 60n
Wilson, Frank J. (The TRANSCOM Coalition: Multi-jurisdictional Issues in ITS), 201n
Wilson, Richard Guy (The Great Civilization), 21n, 29n
Wilson, Woodrow (The Study of Administration), 25n
Winston, Clifford and Shirley, Chad *(Alternate Route: Toward Efficient Urban Transportation)*, 180n
Work of the Public Roads Administration 1946, 45n
Work of the Public Roads Administration 1947, 46n
Works Progress Administration (WPA), 33
World War I, 31
World War II, 32
World Wide Web, impact upon data availability, 207
WPA. *See* Works Progress Administration

Y

The Yellow Book, 46

Z

zero emission vehicle (ZEV), 212–213
ZEV Fact Sheet: Memoranda of Agreement, 213n
Ziegelbauer, J. (Taxes Lurk in New Inflation-Adjusted Treasury Bonds), 90n
Zuckerman, Gregory (Inflation-Linked Bond's Debut Bolsters Treasury department), 90n